ESO ASTROPHYSICS SYMPOSIA
European Southern Observatory

Series Editor: Jacqueline Bergeron

Springer
Berlin
Heidelberg
New York
Barcelona
Hong Kong
London
Milan
Paris
Singapore
Tokyo

ESO ASTROPHYSICS SYMPOSIA
European Southern Observatory

Series Editor: Jacqueline Bergeron

G. Meylan (Ed.), **QSO Absorption Lines**
Proceedings, 1994. XXIII, 471 pages. 1995.

D. Minniti, H.-W. Rix (Eds.), **Spiral Galaxies in the Near-IR**
Proceedings, 1995. X, 350 pages. 1996.

H. U. Käufl, R. Siebenmorgen (Eds.), **The Role of Dust in the Formation of Stars**
Proceedings, 1995. XXII, 461 pages. 1996.

P. A. Shaver (Ed.), **Science with Large Millimetre Arrays**
Proceedings, 1995. XVII, 408 pages. 1996.

J. Bergeron (Ed.), **The Early Universe with the VLT**
Proceedings, 1996. XXII, 438 pages. 1997.

F. Paresce (Ed.), **Science with the VLT Interferometer**
Proceedings, 1996. XXII, 406 pages. 1997.

D. L. Clements, I. Pérez-Fournon (Eds.), **Quasar Hosts**
Proceedings, 1996. XVII, 336 pages. 1997.

L. N. da Costa, A. Renzini (Eds.), **Galaxy Scaling Relations: Origins, Evolution and Applications**
Proceedings, 1996. XX, 404 pages. 1997.

L. Kaper, A. W. Fullerton (Eds.), **Cyclical Variability in Stellar Winds**
Proceedings, 1997. XXII, 415 pages. 1998.

R. Morganti, W. J. Couch (Eds.), **Looking Deep in the Southern Sky**
Proceedings, 1997. XXIII, 336 pages. 1999.

J. R. Walsh, M. R. Rosa (Eds.), **Chemical Evolution from Zero to High Redshift**
Proceedings, 1998. XVIII, 312 pages. 1999.

J. Bergeron, A. Renzini (Eds.), **From Extrasolar Planets to Cosmology:
The VLT Opening Symposium**
Proceedings, 1999. XXVIII, 575 pages. 2000.

A. Weiss, T. G. Abel, V. Hill (Eds.), **The First Stars**
Proceedings, 1999. XIII, 355 pages. 2000.

A. Fitzsimmons, D. Jewitt, R. M. West (Eds.), **Minor Bodies in the Outer Solar System**
Proceedings, 1998. XV, 192 pages. 2000.

L. Kaper, E. P. J. van den Heuvel, P. A. Woudt (Eds.), **Black Holes in Binaries and Galactic Nuclei:
Diagnostics, Demography and Formation**
Proceedings, 1999. XXIII, 378 pages. 2001.

G. Setti, J.-P. Swings (Eds.), **Quasars, AGNs and Related Research Across 2000**
Proceedings, 2000. XVII, 220 pages. 2001.

Series homepage – http://www.springer.de/phys/books/eso/

G. Setti   J.-P. Swings   (Eds.)

Quasars, AGNs and Related Research Across 2000

Conference on the Occasion
of L. Woltjer's 70th Birthday
Held at the Accademia Nazionale
dei Lincei, Rome, Italy 3-5 May 2000

 Springer

Volume Editors

Giancarlo Setti
Dipartimento di Astronomia
Università di Bologna
Via Ranzani 1
40127 Bologna, Italy

Jean-Pierre Swings
Institut d'Astrophysique
Université de Liège
Avenue de Cointe, 5
4000 Liège, Belgium

Series Editor

Jacqueline Bergeron
European Southern Observatory
Karl-Schwarzschild-Strasse 2
85748 Garching, Germany

Library of Congress Cataloging-in-Publication Data applied for.
Die Deutsche Bibliothek - CIP-Einheitsaufnahme

Quasars, AGNs and related research across 2000 : proceedings of the
ESO Workshop, held at Rome, Italy, 3 - 5 May 2000, on the Occasion of
L. Woltjer's 70th Birthday / G. Setti ; J. - P. Swings (ed.). - Berlin
; Heidelberg ; New York ; Barcelona ; Hong Kong ; London ; Milan ;
Paris ; Singapore ; Tokyo : Springer, 2001
  (ESO astrophysics symposia)
  (Physics and astronomy online library)
  ISBN 3-540-42191-2

ISBN 3-540-42191-2 Springer-Verlag Berlin Heidelberg New York

This work is subject to copyright. All rights are reserved, whether the whole or part of the material is concerned, specifically the rights of translation, reprinting, reuse of illustrations, recitation, broadcasting, reproduction on microfilm or in any other way, and storage in data banks. Duplication of this publication or parts thereof is permitted only under the provisions of the German Copyright Law of September 9, 1965, in its current version, and permission for use must always be obtained from Springer-Verlag. Violations are liable for prosecution under the German Copyright Law.

Springer-Verlag Berlin Heidelberg New York
a member of BertelsmannSpringer Science+Business Media GmbH

http://www.springer.de

© Springer-Verlag Berlin Heidelberg 2001
Printed in Germany

The use of general descriptive names, registered names, trademarks, etc. in this publication does not imply, even in the absence of a specific statement, that such names are exempt from the relevant protective laws and regulations and therefore free for general use.

Typesetting: Camera-ready by the authors/editors
Cover design: Erich Kirchner, Heidelberg

Printed on acid-free paper   SPIN: 10719740   55/3141/du - 5 4 3 2 1 0

ACCADEMIA NAZIONALE DEI LINCEI

CENTRO LINCEO INTERDISCIPLINARE
«BENIAMINO SEGRE»

QUASARS, AGNs AND RELATED RESEARCH ACROSS 2000

Conference on the occasion of L. Woltjer's 70th birthday

3-5 MAGGIO 2000

PROGRAMMA - INVITO

ROMA

PALAZZO CORSINI - VIA DELLA LUNGARA, 10

COMITATO SCIENTIFICO

FRANCESCO BERTOLA

MARGHERITA HACK

MARTIN HUBER

GIANCARLO SETTI

JEAN-PIERRE SWINGS

GUSTAV TAMMANN

Home page: www.lincei.it - Posta elettronica anastasi@accademia.lincei.it

PROGRAMMA

Mercoledì 3

14.00 Saluto del Direttore del Centro Linceo Interdisciplinare S. CARRÀ

14.10 M. LONGAIR (Cambridge University): *Challenges in extragalactic astronomy and cosmology for the 21st century*

14.50 G. HASINGER (MPI, Potsdam): *Space distribution of quasars and AGNs: (1) X-ray observations and the XRB*

15.20 M. SCHMIDT (CalTech, Pasadena): *Space distribution of quasars and AGNs: (2) Luminosity functions and evolution*

15.50 Intervallo

16.20 K. KELLERMANN (NRAO, Charlottesville): *Relativistic outflow in quasars and AGN*

16.55 L. WOLTJER (OHP/Arcetri): *The inner parts of quasar host galaxies*

17.25 J. MILLER (Lick Observatory, Santa Cruz): *Keck spectra of the host galaxies of quasars*

17.50 S. D'ODORICO (ESO): *Advances in the study of the IGM at high redshifts with the VLT high resolution spectrograph-UVES*

18.10 P. OSMER (Ohio State University): *The evolution of quasars and their relation to galaxies*

Giovedì 4

9.00 P. SHAVER (ESO): *Why we need larger radiotelescopes: (1) ALMA*

9.25 R. EKERS (ATNF, Epping): *Why we need larger radiotelescopes : (2) SKA*

9.50 N. KARDASHEV (ASC, Moscow): *AGN's central machine physics from space VLBI*

10.15 G. BURBIDGE (UCSD, San Diego): *What is the future for the conflict between real data and consensus theory in extragalactic astronomy*

10.40 Intervallo

11.10 D. MACCHETTO (STScI, Baltimore): *Hunting for massive black holes in galactic nuclei*

11.45 A. CAVALIERE (Università di Roma Tor Vergata): *Massive BH and their galactic environment*

14.00 Y. TANAKA (MPE, Garching): *X-ray studies of AGNs*

14.40 A. MOORWOOD (ESO): *The role of ground based IR astronomy*

15.15 C. CESARSKY (ESO): *The impact of ISO on AGN research*

15.50 Intervallo

16.20 T. COURVOISIER (ISDC, Geneva): *What may we learn from radio to γ-ray monitoring of quasars and AGNs*

16.50 F. PACINI (Arcetri): *Neutron stars in SN remnants*

Venerdì 5

9.00 M. TARENGHI (ESO): *The 8m class telescope and beyond*

9.35 P. LENA (Observatoire de Paris): *The future of near-IR/optical interferometry*

10.10 R. GILMOZZI (ESO): *Science and technology of a 100m telescope: the OWL concept*

10.25 S. BECKWITH (STScI, Baltimore): *The NGST role*

11.00 Intervallo

11.35 R. BONNET (ESA): *The essential role of space astronomy*

12.10 Concluding remarks

Preface

This book contains the proceedings of the international conference held on the premises of the Accademia Nazionale dei Lincei (Rome, May 3–5, 2000) to celebrate the 70th birthday of Lodewijk Woltjer.

The main theme of the conference focussed on the physics, origin and space distribution of AGNs and quasars and their relationship to the environment, a very wide subject which has attracted much of the research interests of Lo over the years. A number of lectures were also dedicated to reviewing the recent observational advancements and those that may be attained by the introduction of new and powerful astronomical instrumentation both from the ground and from space, in recognition of the central role played by Lo in the promotion of the ESO VLT and of his involvement in the shaping of ESA's space programme Horizon 2000+.

The conference was promoted by Martin Huber, Gustav Tammann and ourselves and was attended upon invitation by about 70 participants. We wish to express our gratitude to all participants, friends and colleagues of Lo, who contributed to a very lively and interesting meeting and particularly, of course, to the speakers for their excellent reviews of the various topics and the extra effort of writing them up for the proceedings.

We express our gratitude for the sponsorship of the Interdisciplinary Centre of the Accademia Nazionale dei Lincei, the generous support of ESA toward travel and organization expenses and the publication by ESO of the proceedings. We also wish to warmly thank Denise Caro of the Liège Institute of Astrophysics and Geophysics for her help during the preparation of the conference, Anna Anastasi and collaborators of the Accademia Nazionale dei Lincei for the local organization, Patrizia Braschi of the Institute of Radioastronomy in Bologna for secretarial help and Pamela Bristow of ESO for taking care of the preparation of the proceedings.

Bologna, Liège,
February 2001

Giancarlo Setti
Jean-Pierre Swings

Contents

The Astrophysics and Cosmology of the 21st Century –
Active Galactic Nuclei 1
   M.S. Longair

The X-Ray Background and the Space Distribution of QSOs . 14
   G. Hasinger

Relativistic Outflow in Quasars and AGN 26
   K.I. Kellermann

The Circumnuclear Environment and Early Evolution of AGN 35
   L. Woltjer

Advances in the Study of the IGM at High Redshifts
with the VLT High Resolution Spectrograph UVES 43
   S. D'Odorico

The Evolution of Quasars and Their Relation to Galaxies 50
   P.S. Osmer

Why We Need Larger Radiotelescopes:
The Atacama Large Millimeter Array 56
   P.A. Shaver

AGN's Central Machine Physics from Space VLBI 66
   N.S. Kardashev

Non-Cosmological Redshifts 77
   G. Burbidge

Supermassive Black Holes 96
   F.D. Macchetto

Massive Black Holes in Galactic Nuclei 116
   A. Cavaliere

X-Ray Studies of Active Galactic Nuclei 131
   Y. Tanaka

The Role of Groundbased Infrared Astronomy 143
   A. Moorwood

What May We Learn from Multi-Wavelength Observations
of Active Galactic Nuclei 155
   T.J.-L. Courvoisier

The Relationship Between Supernova Remnants
and Neutron Stars 165
   F. Pacini

The Future of Ground-Based Optical Interferometry 171
   P. Léna

Science and Technology of a 100m Telescope:
the OWL Concept 184
   R. Gilmozzi, P. Dierickx, G. Monnet

The Next Generation Space Telescope 193
   S.V.W. Beckwith

The Essential Role of Space Astronomy 205
   R.M. Bonnet

After Dinner Speech in Honour of Lodewijk Woltjer 217
   R. Lüst

Conference participants, May 4, 2000

List of Participants

Name	Institution
G. Alcaino	Instituto Isaac Newton, Santiago, Chile
N. Baker	Astronomy Dept., Columbia University, New York, USA
H. Balsiger	Physikalisches Institut, University of Berne, Switzerland
S. Beckwith	S.T.Sc.I., Baltimore, USA
J. Bergeron	European Southern Observatory, Garching, Germany
S. Berthet	Federal Office for Education and Science, Berne, Switzerland
F. Bertola	Astronomy Dept., University of Padua, Italy
G. Bignami	Italian Space Agency, Rome, Italy
R. Bonnet	European Space Agency, Paris, France
P. Bouchet	Cerro Tololo, Chile
J. Breysacher	European Southern Observatory, Garching, Germany
G. Burbidge	Dept. of Physics & Center for Astron. and Astrophys., University of California, San Diego, USA
A. Cavaliere	University of Rome "Tor Vergata", Italy
G. Cavallo	European Space Agency, Paris, France
C. Cesarsky	European Southern Observatory, Garching, Germany
C. Chiuderi	Astronomy & Space Sci. Dept., University of Florence, Italy
F. Chiuderi-Drago	Astronomy & Space Sci. Dept., University of Florence, Italy
A. Comastri	Osservatorio Astronomico, Bologna, Italy
G. Contopoulos	Astronomy Center, Academy of Athens, Greece
T. Courvoisier	INTEGRAL Science Data Centre, Versoix, Switzerland
S. D'Odorico	European Southern Observatory, Garching, Germany
J. Danziger	Osservatorio Astronomico, Trieste, Italy

R. Ekers	ATNF/CSIRO, Epping, Australia
D. Enard	European Gravitational Observatory & VIRGO Project, Pisa, Italy
R. Gilmozzi	European Southern Observatory, Chile
H. Habing	Sterrewacht, Leiden University, The Netherlands
G. Hasinger	Astrophysikalisches Institut, Potsdam, Germany
D. Hofstadt	European Southern Observatory, Chile
M. Huber	European Space Agency, Space Science Dept., Noordwijk, The Netherlands
N. Kardashev	Astro Space Center, Lebedev Physical Institute, Moscow, Russia
K. Kellermann	NRAO, Charlottesville, USA
J. Landstreet	Dept. of Physics & Astronomy, University of Western Ontario, Canada
P. Léna	Université Denis Diderot (Paris VII) & Observatoire de Paris, France
P.-O. Lindblad	Stockholm Observatory, Sweden
M. Longair	Cavendish Observatory, University of Cambridge, UK
L. Lucy	Astrophysics Group, Blackett Laboratory, Imperial College, London, UK
R. Lüst	Max-Planck-Institut für Meteorologie, Hamburg, Germany
D. Macchetto	S.T.Sc.I. & ESA, Baltimore, USA
R. Maiolino	Arcetri Astrophysical Observatory, Italy
B. Marano	Dipartimento di Astronomia, University of Bologna, Italy
F. Matteucci	Astronomy Dept., University of Trieste, Italy
J. Miller	Lick Observatory, Santa Cruz, USA
A. Moorwood	European Southern Observatory, Garching, Germany
H. Olthof	ESA/ESTEC, Noordwijk, The Netherlands
J. Ortner	International Space University, Illkirch, France
P. Osmer	Astronomy Dept., Ohio State University, Columbus, USA
F. Pacini	Arcetri Astrophysical Observatory & University of Florence, Italy

G. Palumbo	Italian Space Agency, Rome, Italy
G.C. Perola	Dipartimento di Fisica "E. Amaldi", University of Rome 3, Italy
L. Piro	Istituto di Astrofisica Spaziale CNR, Frascati, Italy
V. Radhakrishnan	Raman Institute, Bangalore, India
M. Rodonò	University of Catania & Astrophysical Observatory, Italy
R. Sancisi	Osservatorio Astronomico, Bologna, Italy
M. Schmidt	California Institute of Technology, Pasadena, USA
G. Setti	Dipartimento di Astronomia, University of Bologna, Italy
P. Shaver	European Southern Observatory, Garching, Germany
R. Sunyaev	Max-Planck-Institut für Astrophysik, Garching, Germany, & Russian Academy of Sciences
J. Surdej	Institut d'Astrophysique, University of Liège, Belgium
J-P. Swings	Institut d'Astrophysique, University of Liège, Belgium
G. Tammann	Astronomisches Institut, Binningen, Switzerland
Y. Tanaka	Max-Planck-Institut für Extraterr. Physik, Garching, Germany, & Institute of Space and Astronautical Science, Japan
M. Tarenghi	European Southern Observatory, Garching, Germany
P. Véron	Observatoire de Haute-Provence, France
M. Véron-Cetty	Observatoire de Haute-Provence, France
R. Wilson	Waaler Str. 29, Rohrbach/Ilm, Germany
L. Woltjer	Observatoire de Haute-Provence, France & Arcetri Astrophysical Observatory, Italy

The Astrophysics and Cosmology of the 21st Century – Active Galactic Nuclei

Malcolm S. Longair

Cavendish Laboratory, Madingley Road, Cambridge CB3 0HE

Abstract. A summary is presented of selected aspects of current understanding of the physics of active galactic nuclei and the problems they present for basic astrophysics. The topics discussed include the evidence for black holes in the centres of galaxies and active galaxies, jets and the role of active galactic nuclei in the origin and evolution of galaxies.

1 Introduction

I am sure that, on the occasion of his 70th birthday, Lo Woltjer would want to take a forward-looking view of how the subject of this meeting, *Active Galaxies and their Nuclei*, will develop in the 21st century. I need not elaborate on Lo's many contributions to the study of active galaxies and high energy astrophysics – suffice to say that his papers are characterised by a deep personal understanding of the physics of astrophysics and, over the years, I have continued to enjoy the profound insights his papers contain and their inspiration for future studies.

In taking a forward-looking approach to the study of active galactic nuclei, I am fully aware of the perils of predicting the future, particularly at the turn of centuries. Just a few examples will illustrate the problem and act as a warning of the fragility of prognostication. I invite the reader to supply the contemporary analogies.

- In 1600, the headlines of the time might have read *Scandal of the Closure of the Greatest Observatory of Modern Times – Director Absconds with 20 Years of Astronomical Data.* Tycho's magnificent data were about to be analysed by Kepler and his three great laws of planetary motion derived from them. These were to pave the way for the Newtonian revolution. No one would have predicted that, within 10 years, the astronomical telescope would have been invented and Galileo would have discovered the satellites of Jupiter, providing a small-scale model of the Copernican system of the World.
- In 1700, the Newtonian revolution was in full swing, but controversy seem to dog Newton at every turn. About 1700, there were still fierce debates about the discovery of the differential calculus, and continental scientists had difficulty in reproducing his *experimentum crucis*, which demonstrated the splitting of light into its "irrefrangible components".

One of my favorite remarks made in the course of the debate by Rizzetti in 1741 reads:

> 'It would be a pretty situation that, in places where experiment is in favour of the law, the prisms for doing it work well, yet in places where it is not in favour, the prisms for doing it work badly.'

- By 1800, Herschel had created his first map of the Universe from star counts in different directions, despite the fact that no distances to any star other than the Sun had been measured. Herschel assumed that all the stars were similar to the Sun and that there was no obscuring matter. In 1802, Thomas Young showed that light consists of waves from his remarkable measurements of the interference of light. More than half a century was to pass before Maxwell showed that light is electromagnetic radiation.
- In 1900, the problems of prediction were even more dramatic. Planck had derived the spectral form for black-body radiation and introduced the 'quantum of action' h, but the derivation of the formula baffled theorists, being at best a garbled version of classical statistical mechanics. He had, however, discovered the key result that quantum concepts are essential in order to understand fundamental physics on the small scale. The null result of the Michelson-Morley experiment continued to baffle the scientific community, although many physicists were homing in on the answer. Although the pointers to the future were there, relativity and quantum mechanics had yet to be discovered.

How much more difficult is it to predict what astrophysics and cosmology will look like in 2020, let alone 2100. All we can say with confidence is that astrophysics and cosmology will advance by continuing to open up new ways of doing astronomy – in all wavebands, we need to increase resolution, both in imaging and spectroscopy, and the sensitivity of detectors. We need to exploit regions of the spectrum in which only the pioneering experiments have been made, for example, in TeV γ-rays. We need to develop the capability of handling very large data sets and comparing them with the predictions of theory. The development of new ways of doing astronomy is essential – gravitational wave astronomy, neutrino astronomy, astroparticle physics and so on. Finally, we need to keep a close eye on developments in cognate sciences, atomic, nuclear, particle and plasma physics as well as chemistry and biology – astrobiology is no longer a dream for the distant future.

I will approach the subject of this meeting from the perspective of three major areas of contemporary astrophysics and ask how the active galaxies, quasars and so on fit into that picture. These are *black holes, jets* and the *origin and evolution of galaxies*. At the same time, I will consider the various facilities for the future which will be needed to advance these studies in the 21st century (for a broader view of astrophysics and cosmology in the 21st century, see Longair 2000).

2 Black Holes

The clear consensus of astrophysicists is that supermassive black holes are the powerhouses responsible for many of the phenomena observed in active galaxies. At the same time, it is essential to take a critical look at the evidence that black holes are indeed present in active galaxies. Of central importance to the story is the evidence for the presence of stellar mass black holes in X-ray binary systems. There are now a good number of examples of Galactic X-ray binary systems in which the masses of the invisible compact X-ray sources are inferred to be greater than $3M_\odot$, a generous upper limit to the masses of neutron stars (Fig. 1, from Charles 1998). In this volume, Tanaka has shown how the X-ray spectral and variability properties of those binaries, which are believed to host stellar mass black holes, are quite distinct and are similar in a number of ways to those of active galactic nuclei. As I will emphasise later, there are further similarities which are of the greatest importance for the physics of black holes in active galactic nuclei.

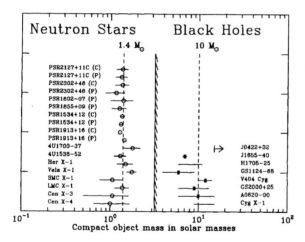

Fig. 1. The mass distributions of neutron stars and black holes. Those systems which are known to possess neutron stars all have masses about $1.4M_\odot$. The masses of the black hole candidates in X-ray binary systems, for which good mass estimates are available, all exceed the upper for neutron star masses of $M \approx 3M_\odot$ (From Charles 1998).

In considering the evidence on the masses of black holes in galaxies, the excellent review of Kormendy and Richstone (1995) is mandatory reading. They describe very clearly the perils of accepting the evidence of high brightness star-like objects and high velocity dispersions in the nuclear regions of galaxies as evidence for black holes. It is essential to find evidence for point-like masses from observations of Keplerian velocity curves in the central regions. It is important to carry out these determinations with very high angular resolution. The determination of the velocity curve in the nearby spiral galaxy NGC 4258 using the intense water maser lines observed by very long baseline interferometry is of special importance. Water maser lines are only observed if the velocity dispersion along the column though which the maser action takes place is less than about 1 km s^{-1}. This means that the masers are either observed in directions tangential to the disc or along the line of sight directly to the nucleus. The resulting velocity curve on the scale of milliarcseconds is shown in Fig. 2.

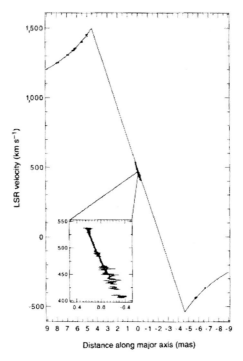

Fig. 2. The rotation curve derived from VLBI observations of water maser sources in the nuclear regions of the nearby spiral galaxy NGC 4258. (Miyoshi *et al.* 1995)

The velocities of the maser sources follow precisely a Keplerian law, $v \propto r^{-1/2}$ at angular distances greater than 4 milliarcsec from the nucleus. The mass within the central region is found to be $M = 3.7 \times 10^7 \, M_\odot$. The masers lie within a distance of 0.2 pc of the nucleus and the corresponding mass density in the central regions must be greater than $4 \times 10^9 M_\odot \, \text{pc}^{-3}$. This value is much greater than the mass density in the densest stellar systems known in our Galaxy, the globular clusters, which have mass density about $10^5 M_\odot \, \text{pc}^{-3}$. The mass density is so great that, even if this mass were made up of neutron stars, they would collide to form a massive black hole in less than cosmological time-scales.

Almost as large mass densities are found in the nucleus of our own Galaxy, but the estimates for the nearest active galaxies, such as M87, are only about $10^6 M_\odot \,\mathrm{pc}^{-3}$. In a number of cases observed by the Hubble Space Telescope, Keplerian rotation curves are observed from the motions of the ionised gas clouds and so there is little doubt that there are high mass densities in the nuclei of many massive galaxies, although determining the masses of the black holes which may be present in the nucleus is far from trivial. These issues are taken up in more detail by Macchetto in this volume.

Generally speaking, wherever the nuclear regions of galaxies have been observed with high spatial and spectroscopic resolution, some of the signatures of compact massive objects have been found and it is quite plausible that all galaxies contain black holes in their nuclei. If this is the case, it raises a host of questions about the relation of the masses of the black holes to other properties of the host galaxies – the ratio of the mass of the black hole to the mass of the bulge and the baryonic mass of the galaxy, the dependence of black hole masses upon morphological type, galaxy environment and so on. With great care, it is now possible to study the galaxies underlying the quasars, as beautifully described in this volume by Miller, but it is not nearly as simple or straightforward to obtain convincing results as might be imagined. The need for very high resolution imaging and spectroscopy of the central regions of normal and active galaxies is obvious.

Fortunately, a completely different approach to the study of black holes in active galaxies has been provided by X-ray observations, particularly the Japanese ASCA satellite. The key to this approach is the discovery of fluorescent lines of iron from a number of Seyfert galaxies which are strong X-ray sources. The origin of the fluorescent lines is the reprocessing of an incident flux of X-rays upon a slab of material at a somewhat lower temperature than that of the X-rays. The ejection of electrons from the K-shell of iron ions in the slab results in the emission of X-ray photons with energy 6.4 keV as the vacancy in the 1s orbit is filled by an electron from the 2p shell. The important realisation has been that the material of an accretion disc about a black hole is a source of fluorescent lines, with the great bonus that the lines are emitted from a deep gravitational potential well in which the accretion disc is rotating at a very high speed about the central black hole. Thus, the 6.4 keV fluorescent iron line acts as a tracer of the velocity field in the accretion disc – the accretion disc extends into regions of the disc in which special and general relativistic effects are large and these can strongly influence the shape of the line profile of the 6.4 keV line.

The best example of this phenomenon to date has been the observation of asymmetric broadening of the 6.4 keV line in the spectrum of the Seyfert 1 galaxy MCG-6-30-15. Fig. 3 shows the spectrum of the 6.4 keV fluorescent line as observed in a long integration by the Japanese X-ray satellite ASCA (Tanaka *et al.* 1995). The profile of this emission line has been determined by subtracting from the overall spectrum a smooth continuum. The key feature

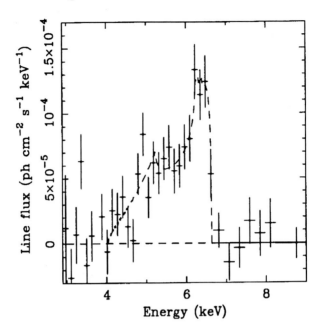

Fig. 3. The broad iron line seen in a long ASCA observation of MCG-6-30-15 (Tanaka *et al.* 1995). The dashed line shows a best fit of the data to the profile expected of an accretion disc about a Schwarzschild black hole. The inclination of the disc is $i = 30°$.

of this remarkable observation is the fact that the spectral line has an abrupt cut-off at energies greater than 6.7 keV, but extends to energies as low as about 4 keV on the low energy side of the line. This type of asymmetry occurs very naturally if the thin accretion disc extends inwards towards the last stable orbit about the black hole. There are two relativistic effects which lead to asymmetries in the line. The first is the gravitational redshift which shifts the spectrum to lower X-ray energies and the second is the transverse Doppler effect which appears in the expression for the Doppler shift of the observed energy of photons emitted from a source moving at an angle θ to the line of sight,

$$\varepsilon_{\rm obs} = \frac{\varepsilon_0}{\gamma \left(1 - \frac{v}{c} \cos \theta \right)}$$

Thus, if the plane of the accretion disc lies at a large angle to the line of sight, $\theta \to \pi/2$, the transverse Doppler shift associated with the Lorenz factor $\gamma = (1 - v^2/c^2)^{-1/2}$ in the denominator dominates and redshifts the line to lower energies. Thus, on an empirical basis, the asymmetric profile of the 6.4 keV line has a natural interpretation if the largest redshifted parts of the line originate from close to the last stable orbit of an accretion disc about a massive black hole, and the disc is observed more or less face-on. This is as might be expected according to the unified picture of active galaxies since, in Seyfert 1 galaxies, the nuclear regions are observed directly and, if the accretion disc is in the same plane as the torus, the plane of the accretion disc would be observed quite close to the plane of the sky. Fig. 3 shows a good fit to the shape of the observed line for the model parameters indicated in the caption. If this interpretation is correct, it means that the largest redshifted emission arises close to the last stable orbit. This type of observation is the most direct evidence we have of emission arising from close to the last stable orbit about a massive black hole.

More details of these observations are described by Tanaka, including the suggestion that, towards the end of the observation, the spectral line profile was even flatter and extended to even softer X-ray energies. This might suggest that the black hole must be a Kerr black hole, since the maximum combined redshift for a Schwarzschild black hole is $1/\sqrt{2}\varepsilon_0 = 4.6$ keV, not so different from the maximum redshift seen in Fig. 3. These are very important observations and it is excellent news that the ESA XMM-Newton X-ray Observatory is reported to have almost completed its commissioning phase. Spectral observations of MCG-6-30-15 are among the priority observations. The significance of such observations for black hole physics is very great and it makes all the more pressing the case for the next generation X-ray spectroscopy mission of ESA, the XEUS Observatory.

3 Jets

One of Lo Woltjer's earliest papers concerned the inverse Compton catastrophe and how it could be avoided if the relativistic beaming of the synchrotron radiation is taken into account. At that time it could not have been predicted the extent to which jets of all sorts play a role in high energy astrophysics. The origin of the relativistic jets needed to power radio galaxies, radio quasars, and microquasars remains a difficult astrophysical problem. They have, however, been put to good use for a variety of astrophysical problems. One of the more remarkable aspects of these studies has been their use in testing orientation-based unification schemes for radio galaxies and radio quasars in the 3CR sample. It turns out that Barthel's original proposal of 1989 that, if the axis of the radio source is observed within an angle of about 45° of the line of sight, a radio quasar is observed and, if at a greater angle, a radio galaxy is observed, seems to pass all the tests that have been made of it. Recently, Tigran Arshakian and I (2000) have reanalysed the problem from the point of view of estimating the mean speeds of advance of the radio 'hot-spots', taking into account environmental, as well as kinematic, effects. Again, the model works very well and the speeds of advance are now in good accord with synchrotron ageing arguments.

One of the consequences of this picture is that the radio galaxies in the 3CR sample provide directly information about the host galaxies of the radio quasars in the sample and some of the types of astrophysics which are likely to be important. A recent example concerns the aligned optical-ultraviolet emission associated with the 3CR radio galaxies at redshifts $z \sim 1$, which were observed by Philip Best, Huub Röttgering and me (2000) as part of our Hubble Space Telescope programme. We discovered that the strength of the optical aligned emission becomes less pronounced, the larger the size of the double radio source. The optical emission from most of the aligned structures in these radio galaxies has now been studied spectroscopically. Using pairs of optical emission lines which are can be used to discriminate between shock and photoionisation, there is a clear distinction among the radio galaxies in the sample (Fig. 4). For those galaxies in which the aligned emission is predominately from luminous, patchy knots with a considerable velocity dispersion, the spectroscopy indicates that the excitation is due to shocks – these are also the smallest double radio sources in the sample. The emission originates from within about 50 kpc of the nucleus of the active galaxy. On the other hand, in the larger radio sources, the emission has a much smaller velocity dispersion, is much less knotty and is excited by photoionisation. These same types of phenomena must play a part in the physics of the 3CR radio quasars as well, if the unification picture is correct.

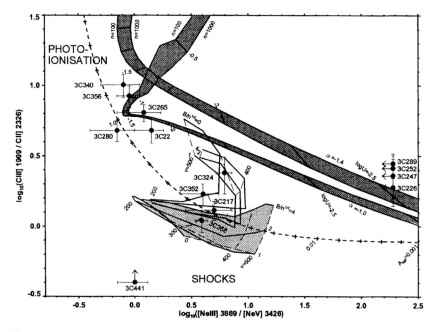

Fig. 4. A line diagnostic diagram for 3CR radio galaxies compared with theoretical predictions. The upper shaded regions correspond to simple photoionisation models, whereas the lower shaded regions correspond to the predictions of shock excitation models (Best, Röttgering and Longair 2000).

Returning to relativistic beaming, the physics of superluminal radio sources is described by Kellermann in this volume and it would be superfluous to repeat many of points he makes which provide insight into the standard relativistic ballistic model. Suffice to say that one of the most intriguing and complex issues is how to generate complete samples of the parent objects which exhibit superluminal motions. The problems of obtaining a clean answer are formidable, because of the difficulty of defining complete high frequency surveys when it is known that the sources are often variable in radio flux density. The statistics of the superluminal sources are of the greatest importance as a test of the standard ballistic model and this will remain a challenge for observers at the US and European VLBI facilities. The study of the detailed physics will be greatly enhanced by the continued progression to higher angular resolution through space VLBI. These capabilities have been clearly demonstrated by the success of the Japanese VSOP-Halca space mission and there are plans for future similar missions such as the NASA ARISE project, which will adopt an inflatable VLBI antenna in space.

The study of superluminal sources has been greatly simulated by the observations of extremely luminous and compact γ-ray sources by the Compton γ-ray Observatory. The remarkable correlation between those sources which exhibit superluminal motions and the most extreme extragalactic γ-ray sources is one of the most exciting discoveries of the CGRO mission – the extreme luminosities are naturally explained by the relativistic beaming of the γ-rays, as well as the radio emission. It is important that the space X-ray and γ-ray observations are complemented by ground-based observations of TeV γ-rays detected by the Cherenkov technique. The long-term monitoring of these sources throughout the radio, X-ray and complete γ-ray wavebands is a major objective of future facilities for high energy astrophysics. Future missions such as INTEGRAL, GLAST and the new generation of ground-based Cherenkov γ-ray experiments are key to these studies.

Another major growth area for these studies will undoubtedly be the study of the Galactic counterparts of the superluminal radio sources, the microquasars. It was a great surprise when Mirabel and Rodriguez showed in 1994 that the Galactic X-ray binary source GRS 1915+105, which contains a stellar mass black hole, exhibits superluminal motion. They demonstrated convincingly that this class of source provides more or less exact replicas of the phenomena observed in radio galaxies and radio quasars, but scaled down by a factor of about one million. The reason for this rather precise scaling is that both the luminosity and the time-scale of variability of the sources are proportional to the mass of the black hole. The only significant difference is that the microquasars are powered by accretion due to the dragging of mass off the primary star of the binary, whereas in the extragalactic case, the mass is presumed to originate by infall and dissipation from the interstellar gas in the parent galaxy. The scaling down by a factor of one million means that phenomena, which would only be observable over periods of decades or centuries in the extragalactic case, can be observed within minutes or hours.

4 Active Galactic Nuclei and the Evolution of Galaxies

Finally, we need to consider the role which active galactic nuclei play in the evolution of galaxies. There has been enormous progress over the last decade in understanding the origin of the large scale distribution of galaxies. The major contributors to this understanding have been observations of the large-scale distribution of galaxies and the power-spectrum of fluctuations in the Cosmic Microwave Background Radiation, together with large-scale numerical simulations of the development of structure according to various realisations of the Cold Dark Matter scenario (for a survey of the reasoning which leads to this conclusion, see Longair (1998)). The upshot of these considerations is that the best physical picture for the origin of galaxies and the large-scale structure of the Universe involves the process of hierarchical clustering, according to which large scale structures are built up by the coalescence of smaller mass objects.

This picture can be confronted by observations of the very distant Universe, in particular, the remarkable observations of the Hubble Deep Field. The dramatic aspects of that image are the very large number of faint blue objects, vastly more than had been predicted by any of the theories, and the much greater percentage of interacting, or coalescing, systems present in that image as compared with that found in the local Universe. These features are in good qualitative agreement with the expectations of the hierarchical picture. In addition, one of the key realisations has been that the global starformation rate can be derived from the integrated ultraviolet luminosity of the star-forming galaxies as a function of cosmic epoch. The diagram derived from observations of the Hubble Deep Field and other estimates at lower redshifts were summarised by Madau and his colleagues, in what has become known as the Madau diagram (Madau *et al.* 1996). This function shows a maximum at redshifts $z \sim 1 - 1.5$.

One of the major concerns about this diagram has been the effects of dust, which is always present where stars are formed and which can strongly attenuate the optical and ultraviolet radiation from distant galaxies. The solution to this problem is to make observations in the submillimetre waveband, since the energy absorbed by the dust heats it to temperatures of about 30-60 K, at which the energy is reradiated in the far-infrared and submillimetre wavebands – the great bonus is that the dust is transparent to radiation in these wavebands and by combining observations in these wavebands with those in the optical waveband a complete census of the star-formation luminosity can be obtained. This has now been undertaken using observations made with the SCUBA submillimetre bolometer array on the JCMT. These observations show that the optical observations seriously underestimate the star-formation rate at large redshifts. A comparison of the optical and submillimetre star formation rates is shown in Fig. 5. It can be seen that the submillimetre observations shift the epoch of maximum star-formation activity to redshifts $z \sim 2 - 3$.

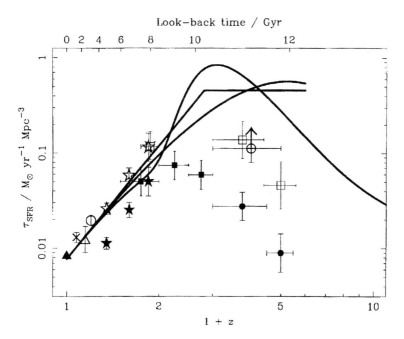

Fig. 5. Comparison of the global star-formation rates in the optical and submillimetre wavebands. The uncorrected optical data are shown by solid dots. The open squares show the optical data corrected for extinction. The open circle is the lower limit to the background from observations of the Hubble Deep Field. The solid lines are estimates of the total star formation rate from a wide variety of data in the far infrared and submillimetre wavebands. It can be seen that the uncorrected optical observations significantly underestimate the star-formation rate because of extinction by interstellar dust (Blain *et al.* 1999).

It is striking that the evolution of the cosmic star-formation rate is similar in form to the evolution of high energy astrophysical activity in the Universe as traced out by the quasars, radio quasars, radio galaxies and faint X-ray sources. Examples of these are presented by Maarten Schmidt in this volume. It would seem remarkable if these phenomena are not intimately related. The maximum in the star-formation rate requires there to be large amounts of gas and dust available and the same gas is also needed to fuel the active galactic nuclei. It is to be hoped that the combination of these different types of study can put the physics of the origin of the strong evolution effects seen in the active galaxies onto a firm astrophysical basis.

There are a number of important issues to be considered in this context. According to the standard Press-Schechter formulation of the process of hierarchical clustering, massive galaxies formed quite late in the Universe and yet we know from the study of the large redshift radio galaxies, that many of these already had stellar masses of the order of $10^{12} M_\odot$ by $z \sim 2$. We also know that many of these active galaxies had to be releasing energy at more or less the Eddington limit at redshifts of 2 and greater. Thus, supermassive black holes must have had time to form by the largest redshifts at which we can observe luminous quasars. This raises a problem concerning the sequence of events which must have taken place at large redshifts when the quasars were first formed. In a plausible picture, the supermassive black holes form as a result of the accretion of baryonic mass and this is limited by the Eddington limiting luminosity. As a result, the time-scale for the doubling of the black hole mass, the Salpeter time-scale, is about 10^8 years. Thus, at the largest redshifts, there may barely be enough time to form supermassive black holes in the nuclei of galaxies.

These studies will undoubtedly be greatly enhanced by the new facilities which are planned for the next generation of astronomical telescopes. In space, the Next Generation Space Telescope will carry on the pioneering work carried out by the HST. On the ground, the Atacama Large Millimetre Array will do for submillimetre studies what the HST has done for optical astronomy. It seems unnecessary to preach the importance of these facilities for the future of astronomy. Lo Woltjer is a shining example to us all of how important it is to push for the facilities we need at the right time and within the right astronomical context. The development of the ESO VLT under his inspired direction shows how necessary it is to keep up the pressure, no matter how obvious the merits of the case might be to the astronomer. We must follow his example to ensure the future health of astrophysics and cosmology. We are all enormously in his debt.

References

1. Arshakian, T.G. and Longair, M.S. (2000). *Mon. Not. R. Astron. Soc.*, **311**, 846.
2. Barthel, P. (1989). *Astrophys. J.*, **336**, 606.
3. Best, P.N., Röttgering, H. and Longair, M.S. (2000). *Mon. Not. R. Astron. Soc.*, **311**, 23.
4. Blain, A.W., Jameson, A., Smail, I., Longair, M.S., Kneib, J.-P.and Ivison, R.J., (1999). *Mon. Not. R. Astron. Soc.*, **309**, 715.
5. Charles, P. (1998). In *Theory of Black Hole Accretion Discs* eds. M.A. Abramowicz, G. Björnsson and J.E. Pringle, 1. Cambridge: Cambridge University Press.
6. Kormendy, J. and Richstone, D.O. (1995). *Ann. Rev. Astron. Astroph.*, **33**, 581.
7. Longair, M.S. (1998). *Galaxy Formation.* Berlin: Springer-Verlag.

8. Longair, M.S. (2000). In *Mathematics Unlimited – 2001 and Beyond*. Berlin: Springer-Verlag (in press).
9. Madau, P., Ferguson, H.C., Dickenson, M.E., Giavalisco, M., Steidel, C.C. & Fruchter, A. (1996). *Mon. Not. R. Astron. Soc.*, **283**, 1388
10. Mirabel, I.F. and Rodriguez, L.F. (1994). *Nature*, **371**, 46.
11. Miyoshi, M., Moran, J., Herrnstein, J., Greenhill,L., Nakai, N., Diamond, P. and Inoue, M. (1995). *Nature*, **373**, 127.
12. Tanaka, Y., Nandra, K., Fabian, A.C., Inoue, H., Otani, C., Dotani, T., Hayashida, K., Iwasawa, K., Kii, T., Kunieda, H., Makino, F. and Matsuoka, M. (1995). *Nature*, **375**, 659.

The X-Ray Background and the Space Distribution of QSOs

Günther Hasinger

Astrophysikalisches Institut Potsdam, An der Sternwarte 16, 14482 Potsdam

Abstract. In this contribution I give a short review of the X-ray background measurements and the understanding of its origin in the light of the 1989 interpretation of G. Setti and L. Woltjer in terms of accretion power of obscured AGN. According to this model, which has been refined and improved in the last decade through deep imaging observations of all X-ray satellites, the X-ray background is mainly the radiation emitted during the buildup-phase of the supermassive black holes which we see in most larger normal galaxies today. The majority of this emission is, however, absorbed by thick gas and dust clouds surrounding the growing black holes.

1 The Cosmic Energy Density Spectrum

In the recent years, the extragalactic background spectrum has been determined with quite high precision over a very broad range of the electromagnetic spectrum (apart from a few inaccessible regions). Figure 1 shows a compilation of some of the most recent determinations of the cosmic energy density spectrum from radio waves to high-energy gamma rays – the "Echos from the Past". Apart from the 2.7 degree K blackbody radiation peak of the CMB, which clearly dominates the energy budget of the universe, three distinct components can be identified in the spectral energy distribution: the Cosmic Infrared Background (CIB[38,16]), the Cosmic Optical Background (COB[37]) and the Cosmic X-ray Background (CXB).

The cosmic X-ray background radiation, a diffuse glow of the whole sky at X-ray wavelengths corresponding to temperatures of several million degrees, was the first extragalactic background emission to be detected [17]. The *ROSAT* satellite has produced the so far most detailed map of the soft X-ray background, which at energies below 1 keV becomes dominated by hot galactic gas in supernova remnants and the interstellar medium.

2 Deep X-Ray Surveys

The X-ray range is one of the few regions of the electromagnetic spectrum, where on one hand the sky emission is dominated by the extragalactic background and on the other hand the modern imaging telescopes are sensitive enough to resolve a substantial fraction of the background into discrete sources. The aim of deep X-ray surveys is to resolve as much of the background as possible, then to obtain reliable optical identifications and redshifts

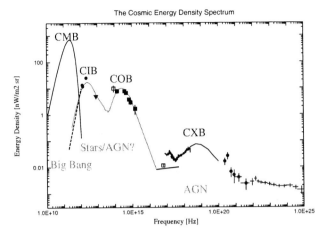

Fig. 1. The cosmic energy density spectrum from radio waves to high energy gamma rays in a νI_ν representation, where a horizontal line corresponds to equal radiation power per decade of energy (from [22]).

of the X-ray sources and finally for all classes of objects to determine luminosity functions and their cosmological evolution. Since the integrated light over cosmic time should correspond exactly to the total X-ray background, this technique provides important additional constraints on the cosmic history.

The deepest survey performed with *ROSAT* is in the direction of the "Lockman Hole", the area on the sky with the absolutely lowest absorption due to interstellar hydrogen. About 1.4 Million seconds observing time has been used for deep X-ray images in the Lockman Hole [21,28] (see Fig. 2a). These images reach a source density of ~ 1000 sources deg^{-2} at a flux limit of 10^{-15} erg cm^{-2} s^{-1}, where 70–80% of the soft X-ray background (0.5–2 keV) has been resolved into discrete sources. Recent deep surveys with *Chandra* and *XMM-Newton* have pushed the sensitivity a factor of five deeper and resolve 80–90% of the soft background into discrete sources with a surface density of ~ 2000 deg^{-2} [34,18,23].

Surveys in the hard X-ray band (2–10 keV) have been pioneered by observations with *ASCA* and *BeppoSAX* [50,9,14], resolving about 30% of the background above 2 keV into discrete sources. The recent *Chandra* and *XMM-Newton* observations have achieved more than an order of magnitude lower sensitivity levels, resolving more than 60% of the hard X-ray background (see Fig. 2b).

The big bottleneck for deep X-ray surveys is the optical identification work. Optical counterparts of the X-ray sources are typically very faint (R<25 for *ROSAT*; R<27 for *Chandra/XMM*), so that excellent X-ray positions and telescopes of the 8–10m class are absolutely necessary. Some optical counterparts are so faint, that even the Keck telescope cannot determine their spectra in a reasonable observing time. In several cases no optical counter-

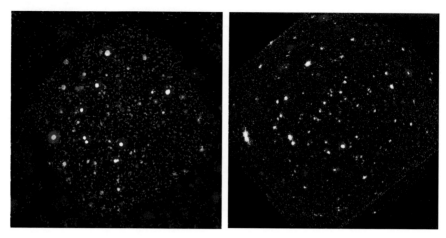

Fig. 2. a: The *ROSAT* Deep HRI Survey in the Lockman Hole[21]; the size of the image is about 35 arcmin across. **b:** The *XMM–Newton* deep survey in the Lockman Hole[23].

part to the X-ray source is found, only on deep near-infrared images some extremely red galaxies (EROs) can be identified in the X-ray error circles [35,27,28].

Those X-ray surveys with a high degree of completeness in optical spectroscopy find predominantly Active Galactic Nuclei (AGN) as counterparts of the faint X-ray source population [7,41,53,3], mainly X-ray and optically unobscured AGN (type-1 Seyferts and QSOs) but also a smaller fraction of obscured AGN (type-2 Seyferts). Spectroscopic identifications of the *BeppoSAX*, *Chandra* and *XMM–Newton* surveys are still far from complete, however a mixture of obscured and unobscured AGN seems to be the dominant population in these samples, too [15,5,18,23].

The *ROSAT* survey in the Lockman Hole is still the deepest optically identified X-ray survey[41,27,28]. Currently 85 out of the 91 X-ray sources are spectroscopically identified. About 80% of those turned out to be active galactic nuclei, among them the highest redshift X-ray selected QSO at z=4.45 [44]. Most AGN are luminous QSOs and Seyfert-1 galaxies with broad emission lines. However about 16 show only narrow emission lines and are classified as Seyfert-2 galaxies. The second most abundant class of objects are groups and clusters of galaxies. Figure 3 shows a correlation between the optical(R)/near-infrared(K') colours and redshift for different types of objects identified in the Lockman Hole in comparison with normal galaxy colours predicted from evolutionary models. The colours of type 1 AGN are relatively blue and scatter widely, while type 2 AGN and cluster galaxies show on average much redder colours, which correlate well with redshift in the region expected for normal galaxies. This indicates that the optical colours of the type 2 AGN are dominated by the light from the host galaxy, probably because the optical nucleus is obscured. Under this assumption one can use photometric techniques to estimate the source redshifts in the range z=1–3. If the redshift estimate is correct, these sources have very high intrinsic X-ray luminosity and might therefore be the first high-redshift examples of luminous obscured AGN, the type-2 QSOs (see [22,28].

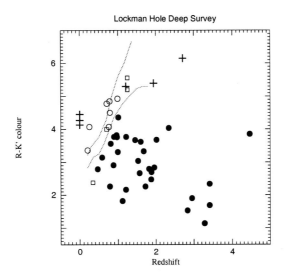

Fig. 3. R–K colours against redshift for X-ray sources in the Lockman Hole marked with different symbols: filled circles are broad-line AGN, open circles are narrow-line AGN (Sy 2), open squares are clusters and groups of galaxies. Plus signs are optical/NIR counterparts without spectroscopic identifications, in 3 cases with photometric redshifts. The dotted lines correspond to unevolved spectral models for E (upper) and S_b (lower) galaxies (from [28].

3 Hard X-Ray Population Synthesis Models

The X-ray background has a significantly harder spectrum than that of local unobscured AGN. This led to the assumption that a large fraction of the background flux is due to obscured AGN, as originally proposed in 1989 by G. Setti and L. Woltjer [39]. Models following the unified AGN schemes, assuming an appropriate mixture of absorbed and unabsorbed AGN spectra folded with cosmological AGN evolution models, can in principle explain the shape of the background spectrum over the whole X-ray range. Apart from the AGN cosmological evolution, the distribution of absorption column densities among different types of AGN and as a function of redshift is one of the major uncertainties. The standard X-ray background population synthesis models assume that the absorption distribution, which has been determined observationally only for local Seyfert galaxies, is independent of X-ray luminosity and redshift. In particular, the fraction of type-2 QSOs should be as high as that of Seyfert-2 galaxies. Also, obscured and unobscured AGNs have been usually assumed to undergo the same cosmological evolution. As shown by Comastri et al. [10], synthesis models can reproduce the hard CXB spectrum up to about 300 keV, together with the total counts and the redshift distributions of AGN seen in medium sensitivity X-ray surveys performed both in the soft and hard X rays. The predictions of the X-ray synthesis models are now starting to be compared with the results of ongoing deep surveys by the *Chandra* and *XMM* satellites[19].

Optical identifications of the faint *Chandra* and *XMM* sources, performed with the VLT and Keck telescopes are expected over the next years to yield a solid statistical basis for an unambiguous population synthesis model of the CXB. In this respect, it could be finally proven that obscured AGN are the main contributors to the hard X-ray background and a population of type-2 QSOs as large as required by the models should be also observed. Other relevant issues like a possible different evolution of obscured and unobscured AGN, and the actual density of AGN at high redshifts should also be settled.

An immediate consequence of the obscured background models is that the radiation produced by accretion processes in AGN emerges completely unabsorbed only at energies well above 10 keV, thus producing the observed maximum of the XRB energy density at 30 keV. Comparison between the background energy density at 30 keV and at 1 keV leads to the suggestion that most (80–90%) of the accretion power in the universe might be absorbed, implying a very large solid angle of the obscuring material as seen from the central source. Fabian et al. [12] suggest that circumnuclear starburst regions are responsible for the large covering factor. They may be both triggering and obscuring most of the nuclear activity.

4 AGN Cosmological Evolution

As discussed above, information about the cosmological evolution of the AGN population is a crucial input into the background synthesis models, but it can not be obtained without taking into account the AGN absorption distribution. However, the AGN X-ray luminosity function in the 0.5–2 keV band has so far mainly been derived ignoring the effects of absorption. First attempts to study AGN cosmological evolution from the *Einstein* Medium Sensitivity Survey (EMSS[11]) or from a combination of medium deep *ROSAT* fields with the EMSS [8] determined the local AGN XLF which has the shape of a broken power law. Their data are consistent with pure luminosity evolution proportional to $(1+z)^{2.7}$ up to a redshift $z_{max} \approx 1.5$, similar to what was found previously in the optical range. This result has been confirmed and improved by more extensive or deeper studies of the AGN XLF, e.g. the RIXOS project[36] or the UK deep survey project[26]. However, these studies were hampered by the uncertain crosscalibration between *Einstein* and *ROSAT*.

In the meantime optical identifications of a large number of *ROSAT* X-ray surveys at various flux limits and solid angle coverage have been completed, comprising the *ROSAT* Bright Survey[43], ROSAT Survey Selected Area North[2], RIXOS[31], the UK deep survey [32] and the Lockman Hole Ultradeep Survey, so that a new AGN soft X-ray luminosity function could be determined, based on *ROSAT* surveys alone[20,33,42]. The log(N)–log(S) function of the overall sample covers six orders of magnitude in flux and agrees within ~ 10% in the overlapping flux regions between different surveys [20].

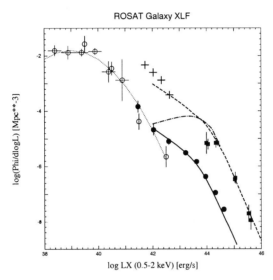

Fig. 4. X-ray (0.5–2 keV) Luminosity functions for AGN and galaxies from *ROSAT* surveys. The local galaxy luminosity function (open circles and dotted line) has been derived in [20]. The AGN luminosity function (from [33]) is shown only for nearby objects ($z < 0.2$, filled circles) and very distant objects ($z > 1.6$, filled squares). Two different luminosity-dependent density evolution models have been fit to the data, one which is close to a pure density evolution model (LDDE2, dashed line) and one where evolution slows down substantially for low luminosities (LDDE1, dash-dotted line). Both models are consistent with all available constraints; their predictions for the density of low luminosity AGN, however, diverge by more than one order of magnitude. The plus signs refer to the optical luminosity function of U-dropout galaxies[37].

Contrary to the previous findings, the new XLF is not consistent with pure luminosity evolution. For the first time we see evidence for strong cosmological evolution of the space density of low-luminosity AGN (e.g. Seyfert galaxy) XLF out to a redshift 1–2, incompatible with pure luminosity evolution. Pure density evolution proportional to $\sim (1+z)^5$ provided a reasonable fit to the *ROSAT* data[20], but overpredicts the total X-ray background, when extrapolated to lower luminosities. Therefore more complicated evolution models have to be taken into account. The latest treatments[42,33] agree that luminosity-dependent density evolution (LDDE) models, where the rate of density evolution is a function of luminosity, can match all constraints. This evolutionary behaviour is similar to the most recently determined optical QSO evolution[52].

Figure 4 compares the luminosity function for local and high-redshift AGN to the luminosity function of local normal and star bursting galaxies. The low-redshift AGN XLF connects smoothly to the galaxy XLF at X-ray luminosities of $L_X \approx 10^{42}$ erg/s. Around this luminosity there is some ambiguity about the relative contribution between the nuclear AGN light and diffuse galactic X-ray emission processes. For clarity, measurements of the high-redshift AGN XLF are only shown for the two highest redshift shells ($1.6 < z < 2.3$ and $2.3 < z < 4.5$) from the data of Miyaji et al. [33]. The apparent deficiency in the lowest luminosity bins is most likely due to incompleteness. Two luminosity-dependent density evolution models are shown, which fit all observational constrains well: the LDDE1 model (dash-dotted line), which is similar to the LDDE model of Schmidt et al. [42], has a rapid slow-down of the density evolution below X-ray luminosities of 10^{44} erg/s and produces $\sim 60\%$ of the extragalactic 0.5–2 keV background. The LDDE2 model (dashed line) is not very much different from a pure density evolution model and produces $\sim 90\%$ of the soft background. The constraints for the XLF of faint, high-redshift Seyfert galaxies, which can produce a significant fraction of the soft X-ray background and, depending on absorption properties, an even larger fraction of the hard X-ray background, are therefore still quite uncertain (the range is a factor of ~ 25 at $\log L_X = 42$). The LDDE2 model predicts a high-redshift AGN space density which is close to that of local normal galaxies just above the break of the luminosity function and to that of high-redshift galaxies selected as U-dropouts[37]. On the contrary, the LDDE1 model predicts a dearth of high-redshift Seyfert galaxies. A choice between these two possibilities will soon be possible with the even deeper X-ray surveys being performed with the *Chandra* and *XMM* observatories.

5 The Space Density of High-Redshift QSO

The X-ray data can also give important new information on the AGN evolution at very high redshifts and therefore on the epoch of black hole formation and the accretion history in the early universe. It is well known from optical samples that the strong evolution of the space density of high luminosity QSOs slows down beyond a redshift of ~ 1.5 and that the space density decreases significantly beyond $z \approx 2.7$ (see [40]). Different selection techniques have to be used below and above a redshift of ~ 2.2 leading to possible

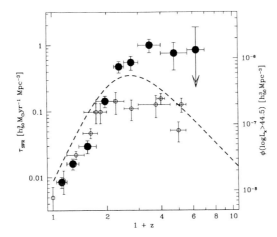

Fig. 5. Cosmic star formation history τ_{SFR} (left Y-axis) compared to the space density ϕ of luminous X-ray selected QSOs (right Y-axis). Filled circles give the comoving number density of *ROSAT* QSOs with $\log L_X > 44.5$ erg/s (from [33]. Open circles give the extinction-corrected optical/UV measurements of the star formation rate by Steidel et al. [49]. The dashed line indicates the simplest star formation history model by Blain et al. (2000, priv. comm.), which explains the whole FIR/sub-mm background light by dusty star formation.

systematic uncertainties in the optical data. Radio-selected QSOs, however, confirm the decline at high redshift and indicate that it is apparently not due to an increase in dust obscuration[46]. The *ROSAT* sample of QSOs allows now for the first time to determine the space density of X-ray selected AGN in the whole range of $0 < z < 5$ with one technique. Fig. 5 (right scale) shows the space density of AGN with X-ray luminosity $\log L_X > 44.5$ erg/s as a function of redshift. Above this luminosity the XLF is a steep power law for all observed redshift shells and the data in Fig. 5 have been determined by assuming the slope to be independent of redshift and fitting the normalization of the XLF[33]. The X-ray data do not show a significant decrease of the space density of high-redshift, high-luminosity X-ray selected AGN and appear to be marginally inconsistent with the optical and radio determinations[33]. However, the X-ray surveys still suffer from small sample sizes at high redshift, so that significantly larger solid angles have to be covered to a similar depth and optical completeness as the ultradeep HRI survey in order to get a clear picture of the AGN density at high redshift. The array of on-going *Chandra* and *XMM* surveys in other fields than the Lockman Hole will be of great help in this respect.

6 AGN Contribution to the Star Formation History

The star formation history in the universe out to redshifts of four has been studied in the last few years by optical and NIR observations using ground-based telescopes and deep photometric surveys with the Hubble Space Telescope (see e.g. [6] for a review). The open circles in Fig. 5 show the compilation of the most recent observational determinations of the optical/UV star forming rate (SFR)[49] (left scale), which suggest that the star formation rate was constant beyond a redshift around 1.5. These data points, however, have to be regarded as lower limits of the true SFR because some of the light emitted in star bursts can be significantly obscured. Deep SCUBA surveys have detected a population of optically faint galaxies, luminous in the sub-mm band, which could produce a significant fraction of the CIB signal [47,25,4]. If all of the CIB should be due to star forming processes, a large population of strongly obscured star bursting galaxies would be missing from the optical/UV surveys at high redshifts. The dashed line in figure 5 sketches one of the SFR models by Blain et al., that is able to produce all of the CIB by early star formation.

It is interesting to note that the star formation history, at least at redshifts below 2, has a dependence on redshift which is very similar to the cosmological evolution of the AGN space density (see Fig. 5). Could it be, that active galactic nuclei contribute significantly to the faint sub-mm source population? The obscured background synthesis models has important consequences for this question. The absorbed AGN will suffer severe extinction and therefore, unlike classical QSOs, would not be prominent at optical wavelengths. The light of the optical counterparts of distant absorbed X-ray sources should therefore be dominated by the host galaxy, which indeed seems to be the case for a significant number of faint X-ray sources. If most of the accretion power is being absorbed by gas and dust, it will have to be reradiated in the FIR range and be redshifted into the sub-mm band. AGN could therefore contribute a substantial fraction to the cosmic FIR/sub-mm background which has already partly been resolved by deep SCUBA surveys. The background population synthesis models have recently been used by Almaini et al. [1] to predict the AGN contribution to the sub-mm background and source counts. Depending on the assumptions about cosmology and AGN space density at high redshifts they predict that a significant fraction of the sub-mm source counts at the current SCUBA flux limit could be associated with active galactic nuclei. Interestingly, some of the first optical identifications of SCUBA sources indicate a significant AGN contribution[48]. The first joint deep X-ray/sub-mm survey in the Hubble Deep Field North[24,45], on the other hand, did not find any common X-ray/sub-mm sources.

Another, largely independent line of arguments leads to the conclusion that accretion processes may produce an important contribution to the extragalactic background light. Dynamical studies[30] come to the conclusion that massive dark objects, most likely dormant black holes, are ubiquitous in nearby galaxies. There is a correlation between the black hole mass and the bulge mass of a galaxy: $M_{BH} \approx 6 \times 10^{-3} M_{Bulge}$. Since gravitational energy release through standard accretion of matter onto a black hole is producing radiation about 100 times more efficiently than the thermonuclear fusion processes in stars, the total amount of light produced by accretion in the universe should be of the same order of magnitude as that produced by stars. A more detailed treatment following this argument comes to the conclusion that the AGN contribution should be about 1/5 of the stellar light in the universe[13]. The AGN background synthesis models can therefore also explain the mass distribution of dark remnant black holes in the centers of nearby galaxies by conventional accretion which is largely hidden by obscuration.

Regardless of whether the FIR light of AGN is from dust heated by stellar processes or by accretion onto the massive black hole, these studies indicate that a large contribution to the light emission history in the early universe could come from sources associated with AGN, which are most easily pinpointed by sensitive X-ray observations. Future FIR/X-ray surveys, deeper and wider than those available today, will therefore be a very powerful tool to disentangle the different processes dominating the universe in the redshift range $2 < z < 5$.

7 NGC 6240 – Representative of the Obscured Universe?

The ultraluminous IRAS galaxy NGC 6420 is a starburst galaxy with a double nucleus with some evidence for a weak active nucleus. Its position in the mid-infrared diagnostic diagram based on ISO spectroscopy[29] indicates that its bolometric luminosity, which is dominated by the far-infrared dust emission, should be mainly heated by star formation processes. Recently, however, *BeppoSAX* hard X-ray spectroscopy revealed an underlying very luminous, though heavily obscured active galactic nucleus, which could well be responsible for the major fraction of the bolometric luminosity[51]. Fig. 6 shows the spectral energy distribution (SED) of NGC 6240 in a νI_ν representation from the radio to the hard X-ray band. The SED peaks in the far-infrared dust band and shows the mid-infrared polycyclic aromatic hydrocarbonic emission bands as well as the synchrotron radio emission typical for starburst galaxies. The optical starlight peak is lower than the dust emission peak – similar to the overall extragalactic energy distribution. Compared to the average SED of unobscured AGN, the hard X-ray nonthermal AGN continuum emission of NGC 6240 is severely absorbed below 15 keV and a strong reprocessed iron line is seen. The UV to soft X-ray nuclear continuum is completely obscured.

The SEDs of three AGN from the *ROSAT* deep survey in the Lockman Hole, which have been observed over a broad wavelength range from radio to X-rays have been superposed on this diagram. For a comparison on the same absolute scale the Lockman SEDs have been shifted to the redshift of

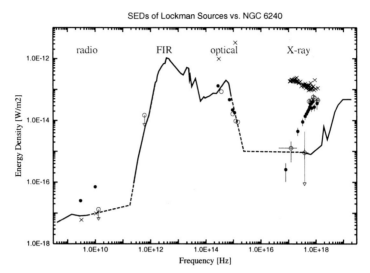

Fig. 6. Spectral energy distribution of the ultraluminous IRAS galaxy (ULIRG) NGC 6240 from radio to hard X-ray frequencies (thick solid line, dashed in unobserved portions), compared to the spectral energy distribution of three high-redshift X-ray sources from the *ROSAT* Deep Survey: 12A (filled circles), 14Z (open circles) and 32A (crosses). The latter SEDs have been normalised to the redshift of NGC 6240. From [22].

NGC 6240 and scaled by the square of their luminosity distances. One of the *ROSAT* deep survey sources (32A) is a broad-line QSO and its SED fits nicely to the average AGN SED. The other two sources are type-2 (absorbed) Seyfert galaxies or quasars and their SED is surprisingly similar to that of NGC 6240, albeit with lower absorption columns, so that the X-ray continuum starts to be obscured only below 3 keV (in the restframe of the objects).

Comparing Fig. 6 with Fig.1 it is easily understandable, how the shape of the X-ray background can be produced by a superposition of AGN spectra with different amounts of absorption. This figure is a nice illustration of the potential relevance of the obscured AGN X-ray background model for the the FIR/sub-mm background, since it shows that X-ray-selected high-redshift AGN can indeed have SEDs similar to local ULIRGS.

References

1. Almaini O., Lawrence A., Boyle B., 1999, MNRAS 305, L59
2. Appenzeller I. et al., 1998, ApJS 117, 319
3. Akiyama M. et al., 2000, ApJ 532, 700
4. Barger A.J. et al., 1998, Nature 394, 248
5. Barger A.J. et al., 2000, AJ (in press) [astro-ph/0007175]
6. Blain A.W. et al., 1999, MNRAS 302, 632 (Erratum MNRAS 307, 480)

7. Bower R. et al., 1996, MNRAS 281, 59
8. Boyle B.J. et al., 1994, MNRAS 260, 49
9. Cagnoni I., Della Ceca R. and Maccacaro T. 1998, ApJ 493, 54
10. Comastri A., Setti G., Zamorani G. and Hasinger G. 1995, A&A, 296, 1
11. Della Ceca R. et al., 1992, ApJ 389, 491
12. Fabian A.C. et al., 1998, MNRAS 297, L11
13. Fabian A.C., Iwasawa K., 1999, MNRAS 303, L34
14. Fiore F. et al., 1999, MNRAS 306, L55
15. Fiore F. et al., 2000 [astro–ph/0003273]
16. Fixsen D.J. et al., 1998, ApJ 508, 123
17. Giacconi R. et al., 1962, Phys.Rev.Letters 9, 439
18. Giacconi R. et al., 2000, ApJ (submitted) [astro–ph/0007240]
19. Gilli R., Salvati M., Hasinger G., 2000, A&A (in press) [astro–ph/0011341]
20. Hasinger G., 1998, AN 319, 37
21. Hasinger G. et al., 1998, A&A, 329, 482
22. Hasinger G., 2000, Lecture Notes in Physics; 548, p.433 [astro–ph/0001360]
23. Hasinger G., et al., 2000, A&A (in press) [astro–ph/0011271]
24. Hornschemeier A.E. et al., 2000, ApJ 541, 49
25. Hughes D.H. et al., 1998, Nat 394, 241
26. Jones L.R. et al., 1997, MNRAS 285, 547
27. Lehmann I. et al., 2000, A&A 354, 35
28. Lehmann I. et al., 2001, A&A (in prep.)
29. Lutz D., et al., 1996, A&A 315, L137
30. Magorrian J. et al., 1998, AJ 115, 2285
31. Mason K.O. et al., 2000, MNRAS 311, 456
32. McHardy I.M. et al., 1998, MNRAS 295, 641
33. Miyaji T., Hasinger G., Schmidt M., 2000, A&A 353, 25
34. Mushotzky R.F., et al., Nature 404, 459
35. Newsam A.M. et al. 1997, MNRAS 292, 378
36. Page M.J. et al., 1996, MNRAS 281, 579
37. Pozzetti L. et al., 1998, MNRAS 298, 1133
38. Puget J.-L. et al., 1996, A&A 308, L5
39. Setti G. and Woltjer L. 1989, A&A 224, L21
40. Schmidt M., Schneider D.P., Gunn J.E., 1995, AJ 110 68
41. Schmidt M. et al. 1998, A&A, 329, 495
42. Schmidt M. et al. 1999, MPE report 272, 213 [astro–ph/9908295]
43. Schwope A. et al., 2000, AN 321, 1
44. Schneider D. et al., 1998, AJ 115, 1230
45. Severgnini P., et al., 2000, A&A 360, 457
46. Shaver P.A. et al., 1999, in Highly Redshifted Radio Lines eds. Carilli C. et al. (PASP: San Francisco), 163 [astro–ph/9801211]
47. Smail I., Ivison R.J., Blain A.W., 1997, ApJ 490, L5
48. Smail, I., Ivison, R.J., Blain, A.W., Kneib, J.-P., in *After the Dark Ages: When Galaxies were Young*, eds. Holt, S., Smith, E.P., AIP, (1999) [astro–ph/9810281]
49. Steidel C.C. et al., 1999, ApJ 519, 1
50. Ueda Y. et al., 1998, Nature 391, 866
51. Vignati P., et al., 1999, A&A 349, 57
52. Wisotzki L., 1998, AN 319, 257
53. Zamorani G. et al., 1999, A&A 346, 731

Relativistic Outflow in Quasars and AGN

K.I. Kellermann

National Radio Astronomy Observatory, Charlottesville, VA 22901, U.S.A.

Abstract. Very long baseline interferometer measurements provide direct evidence for highly collimated relativistic motion in quasars and AGN with apparent velocity typically around 3c, but extending up to about 20c. In general, radio galaxy jets appear to move with slower speed than quasar jets, while jets associated with quasars that show strong gamma-ray emission appear to move the fastest. Surprisingly, there appears to be little or no evidence for Doppler boosting at the level corresponding to the observed superluminal motion. It is difficult to explain the high brightness temperatures deduced from observations of intra-day-variability in terms of Doppler boosting. Coherent emission mechanisms may need to be considered.

1 Early Evidence of Relativistic Motion

Probably the first evidence for highly directed relativistic motion or anisotropic emission came from the discovery in the mid 1960's of the rapid variability in some quasars and active galactic nuclei [26,6]. These early observations showed evidence of changes by as much as 25 percent over a few weeks as well as apparently significant day to day variations [22]. It was realized by many, particularly Hoyle et al. [11], that the observed rapid variability implied such small linear dimensions, that if the quasar redshifts were interpreted as a measure of distance, the radio sources would be rapidly extinguished by inverse Compton cooling. Later, Kellermann and Pauliny-Toth [15] put these arguments on a quantitative observational basis showing that the maximum sustainable brightness temperature for an incoherent source of synchrotron radiation is about 10^{12} K.

However, Lo Woltjer [34] pointed out that if the relativistic electrons preferentially move along the magnetic field lines, the radiation is beamed along the direction of motion, and the cross-section for inverse Compton scattering is greatly reduced. With great perception Woltjer remarked that

> If the relativistic electrons and the photons they emit, both move nearly parallel to the line of sight, the time scale of the variations in emission can be much shorter than the size of the region divided by the velocity of light.

With great foresight he went on to say

> Since the emission process is strongly anisotropic, there is no reason to expect that the overall emission from the QSO to be isotropic.

Woltjer did not specifically point out that because of the finite light travel time, if you could actually see the relativistic plasma cloud move, the apparent transverse velocity would exceed the speed of light. The following year, Martin Rees [24,25] discussed essentially the same effect, but with a somewhat different formulation. In 1969, VLBI observations [33,4] showed actual evidence for the predicted high velocity outflow with apparent velocities typically of the order of ten times the speed of light. But, the observations were somewhat indirect, and one had to have faith in the interpretation of the very limited interferometer data which was not completely unambiguous [7].

2 Parsec Scale Structure of Quasars and AGN

Today using the VLBA and global VLBI networks we can image the radio emission from quasars and AGN on milliarcsecond scales corresponding to linear dimensions of a few tens of parsecs at cosmological distances and as small as a few light days for nearby AGN. We see directly the relativistic flow of plasma along well collimated jets. Typically the jets are one-sided, but in some sources, particularly those associated with AGN such as NGC 1052, the structure is more symmetric [14]. Usually the jets are well collimated and are unresolved in the direction transverse to their flow, although there may be significant curvature, especially close to the nucleus.

In some cases, however, instead of a well-collimated jet, there is a broad plume, or more complex two dimensional structure. In the case of M87, the well known jet appears to bifurcate only about 0.1 parsecs from the nucleus.

In general, the parsec scale jets appear well aligned with the more extensive arcsecond (kiloparsec) scale jets of the type observed with the VLA or Westerbork Array. This implies that the kiloparsec jets are initially collimated on scales of a few parsecs or less on time scales of a few years, or less, and remain focused for times up to at least a million years.

In some radio galaxies with symmetric two sided structure, the appearance of the source is very frequency dependent suggesting free-free absorption from a thin disk or torus, probably associated with the accretion disk surrounding the central engine. Observations of this type provide a unique means to directly study the accretion on to a massive black hole from the surrounding disk or torus where the density of the ionized material is of the order of 10^{4-5} cm^{-3} within one parsec from the black hole (e.g., [32]). This is comparable to the density of the optical broad-line emission region suggesting that the two may be identical.

3 Kinematics

Repeated observations of the compact radio sources in quasars and AGN directly show the outflow away from the AGN as predicted by Woltjer more than 30 years ago. Current research efforts are directed toward describing the kinematics of the relativistic flow. For the past five years, Rene Vermeulen, Tony Zensus, Marshall Cohen, and I have been making regular observations of a large sample of quasars and AGN. We want to know:

> Where does the relativistic flow get accelerated and collimated to form jets?
>
> Does the flow follow curved or straight trajectories?
>
> Do different components within a jet follow the same or different trajectories? Do they have the same or different speeds?
>
> Are there accelerations or decelerations?
>
> Is the apparent velocity related to luminosity, x-ray or gamma ray emission? Or to anything else?
>
> Is the time of appearance of a new component emerging from the nucleus related to the start of a flux density outburst?
>
> Are there any observed differences in the kinematics of radio galaxies, quasars, or BL Lac Objects?
>
> Are the simple ballistic models correct, or are there differences between the bulk flow velocity and the pattern velocity as might be expected if the observed motions are due to the propagation of shocks rather than the flow of material (e.g., [30])?

Some preliminary results of our observations have already been published [16,17]. It is perhaps curious, that the observed flow is always outward, away from the apparent central engine, yet it is widely thought that the source of energy is *infall* onto a massive black hole. Often the trajectory is curved, especially close to the nucleus. Generally, each component moves with constant velocity until it fades away, and each component appears to move along a similar trajectory. However, in some carefully studied sources, such as 3C 345, there is evidence for small accelerations and slightly different trajectories (e.g., [20]).

4 Relativistic Beaming

The observed motions are generally interpreted in terms of relativistic beaming models (e.g., [2]) and in particular, unified models which assume that the intrinsic velocity of ejection of the jet components is close to the speed of light. The differences observed in the properties of quasars, AGN, and BL Lac Objects, particularly the apparent velocity, is thought to be primarily due to the orientation of the relativistic beam with respect to the line of sight. The goal is to understand the process of acceleration and collimation of relativistic jets which might lead to better insight into the nature of the central engine.

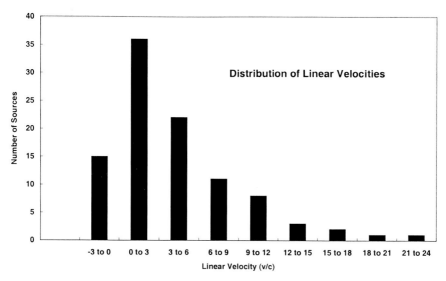

Fig. 1. The distribution of apparent velocity for all sources in which there has been a reliable measure of component motion. The median observed apparent velocity is about 3c, which is considerably smaller than most published observations of superluminal motions (e.g. [30])

The apparent transverse velocity of separation of the jet components from the nucleus is given by

$$(v/c)_{app} = \beta_p \sin(\theta)/(1 - \beta_p \cos\theta)$$

where $\beta_p = v/c$ is the pattern velocity describing the motion of components along the jet. The apparent transverse velocity reaches a maximum value of γ_c at an angle $\theta \sim 1/\gamma$, where γ is the Lorentz factor defined by $\gamma = (1-\beta^2)^{-1/2}$. But, in a randomly oriented sample, most jets will be aligned close to the plane of the sky, and the apparent projected velocity will be close to the speed of light as foreshortening effects will be small. The a-priori probability of observing a source within an angle θ to the line of sight is of the order $1/\theta^2$ corresponding to a probability of $1/\gamma^2$ for observing an apparent velocity γc.

Early observations suggested that about half of all known compact radio sources showed superluminal motion with a surprisingly large value of $(v/c)_{app}$ of 5 to 10 [5]. It was quickly realized, however, that due to relativistic beaming, the apparent flux density, S, of a moving component is enhanced over its stationary value S_0, by an amount

$$S/S_0 = \delta^{3+\alpha}$$

where $\delta = \gamma^{-1}(1 - \beta_b \cos\theta)^{-1}$ is the Doppler factor, and where β_b is the bulk velocity of the relativistic flow.

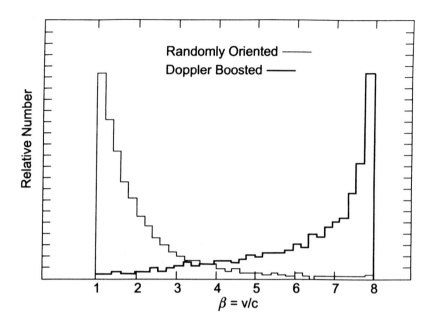

Fig. 2. Distribution of apparent velocity for an intrinsic value of $\beta = 0.99$ ($\gamma = 8$) and random orientation of motion for the case where there is no Doppler bias and the observed distribution of velocity reflects only the solid angle distribution (thin line), and where Doppler boosting biases the observed distribution toward motion oriented close to the line of sight and the most probable velocity is $\sim \gamma c$ (thick line). (Adapted from Vermeulen [29])

Provided that $\beta_p = \beta_b$, observations of the apparent component velocity and the ratio of flux densities of approaching and receding components can, in principle, be used to solve uniquely for β and θ. In this case, in a flux density limited sample, the combined effect of solid angle selection and Doppler boosting leads to a preferential apparent velocity close to γc [30]. On the other hand, if Doppler boosting is not important, then $(v/c)_{app} \sim 1$. Figure 2 shows the expected distribution of apparent velocities for the case where the $\gamma_p = \gamma_b = 8$ ($\beta_p = \beta_b = 0.99$).

How do we reconcile the difference between the observed and predicted distributions? Possibly, there is a spread in intrinsic velocity (e.g., [19]). Alternately, the bulk flow velocity may be much less than the pattern velocity, so that there is relatively little Doppler bias in favor of observing beams which are oriented close to the critical angle. Indeed, the observed distribution of apparent velocity is not unlike the simple *light echo model* discussed by Ekers and Laing [8] where there is no Doppler bias, and the observed distribution depends only on orientation.

There also appears to be a correlation between apparent velocity and luminosity. As might be expected if the apparent luminosity depends more on geometry than on intrinsic properties, the weakest sources, which are primarily associated with nearby AGN rather than quasars, show the smallest velocities. For the high luminosity sources, (L $\geq 10^{25}$ W/Hz) however, there appears to be no significant difference in apparent velocity of quasars, AGN, or BL Lac Objects. This is perhaps curious, as the unified models assume a different orientation for these different classes of extragalactic radio source.

For those sources which have been identified as strong gamma-ray emitters [9] the mean value of $(v/c)_{app}$ is 5.7 ± 0.6 compared with a value of 3.9 ± 0.4 for non gamma-ray sources. This is consistent with the idea that gamma-ray sources are more highly beamed along the line of sight, but it must be remembered that there is no clearly defined class of gamma-ray-loud and gamma-ray-quiet objects, unlike the well established radio-loud and radio-quiet dichotomy. Currently, the range of observable gamma-ray luminosity is limited to the strongest few dozen sources. More sensitive observations with the next generation of gamma-ray observatories, such as GLAST, will be of great interest.

5 The Brightness Temperature Problem

The effect of inverse Compton scattering is usually expressed as an upper limit on the brightness temperature. Under equilibrium conditions, the maximum brightness temperature of a stationary incoherent source of synchrotron radiation from relativistic electrons is about 10^{12} K [15]. Readhead [23] has argued that if there is equipartition between the magnetic energy density and relativistic particle density, then the peak brightness temperatures are nearly an order of magnitude lower. The concept of equilibrium between particle and magnetic energies was introduced long ago to minimize the apparent excessive energy content in extended radio galaxies with characteristic ages of 10^{8-9} years [3]. However, it seems unlikely that equipartition conditions will exist in sources where there is clearly a sudden highly collimated violent release of relativistic plasma. More likely, in these sources with typical ages of only 1 to 100 years estimated from variability observations, there is likely to be an excess of particle over magnetic energy.

What are the observational constraints to the brightness temperature? Ground based observations with the VLBA suggest apparent maximum brightness temperatures of 10^{11} to 10^{12} K or just in the range expected from the inverse Compton limit. However, the maximum measurable brightness temperature depends only on the flux density and interferometer baseline length, independent of wavelength. For a typical flux density of a few Janskys and baseline lengths of 10,000 km, this corresponds to about 10^{12} K so that it is not possible to obtain a stringent test of the inverse Compton limit using interferometer baselines on the surface of the earth.

Very Long Baseline Interferometer observations using orbiting radio telescopes in conjunction with ground radio telescopes can increase the brightness temperature limit beyond that which is obtained using terrestrial baselines alone. Observations made with the NASA TDRSS satellite [18] and the Japanese space VLBI satellite, HALCA [10], in conjunction with the VLBA and other ground-based arrays have suggested rest frame brightness temperatures in excess of the 10^{12} K inverse Compton limit.

The highest observed brightness temperatures come not from direct interferometer measurements, but from the observations of very rapid time variations on the order of hours, the so-called Intra-Day-Variability (IDV). Using the same light travel time arguments that were used in the 1960s following the discovery of the first radio flux density variations, brightness temperatures as high as 10^{21} K [13] were derived for the most rapid variables which shows changes time scales less than one hour.

These light-travel time arguments are, of course, valid only if the observed variations are intrinsic and not due to propagation effects. The apparent correlation of radio and optical variations argue in favor of this interpretation [31] which appeared to pose difficult constraints on conventional synchrotron models. However, recent observations provide convincing evidence that the most rapid variability is due to interstellar scintillations and is not intrinsic. Simultaneous observations made with the VLA and Australia Telescope, located 10,000 km apart, showed a time delay of about two minutes in the observed light curve of the quasar PKS 0405-385. This corresponds to a velocity of motion of irregularities in the interstellar medium of about 75 km/sec and provides further support for an interpretation in terms of propagation effects [12].

Appealing to interstellar scintillations mitigates, but does not eliminate the puzzle of understanding the apparent excess brightness temperatures. While interpretation in terms of intrinsic variations imply nanosecond sizes and $T_B \sim 10^{20}$ K, only sources smaller than about ten microarcseconds will scintillate due to irregularities in the interstellar medium. This corresponds to brightness temperatures in excess of 10^{14} K, still well above the inverse Compton limit.

How do we explain these apparent excessive brightness temperatures?

a) The most common interpretation invokes the effect of relativistic beaming which will enhance the apparent brightness temperature by the Doppler factor

$$\delta = \gamma^{-1}(1 - \cos\theta)^{-1} .$$

However, this would require Doppler factors in excess of 100 to account for the high brightness temperatures deduced from observations of time variability. There is no evidence from observations of component motions that such large Doppler factors exist.

b) The limit of 10^{12} K applies only to incoherent synchrotron radiation. Various authors have discussed different coherent processes, including coherent plasma waves and stimulated synchrotron emission (synchrotron masers). A necessary condition for maser activity or negative absorption is the presence of an inverted electron energy distribution [28] but there has been no consensus whether or not synchrotron maser activity can occur in practice e.g., [1,21,35,36].

c) It must be appreciated that the 10^{12} K limit is an equilibrium value and that it is possible to maintain a larger brightness temperature for limited periods if the initial injection of relativistic electrons extends to sufficiently large energy [15]. The lifetime due to first order inverse Compton is easy to calculate as it is mathematically equivalent to synchrotron radiation loss (e.g., [27]). For example at 10 GHz, T_B can remain greater than 10^{15} K for as long as a month and can exceed 10^{12} K for up to 100 years. Calculating the lifetime including the effects of higher order scattering taking into account the small K-N cross-section for high order scattering as well as possible replenishment of the electron population due to pair production is more complex.

Acknowledgments

I thank my colleagues Rene Vermeulen, Anton Zensus, and Marshall Cohen who have collaborated on much of the research reported here.

References

1. Bekefi, G. & Brown, S.C. (1961) Am. J. Phys., 29, 404
2. Blandford, R. & Konigl, A. (1979) ApJ, 232, 34
3. Burbidge, G. (1959) ApJ, 129, 849
4. Cohen, M. et al. (1971) ApJ, 170, 207
5. Cohen, M.H. et al. (1977) Nature, 268, 405
6. Dent, W. (1965) Science, 148, 1458
7. Dent, W. (1972) Science, 175, 1105
8. Ekers, R. & Laing, H. (1990) in Parsec Scale Radio Jets, eds. J. Zensus & T. Pearson, (San Francisco: Astron. Soc. Pacific), 333
9. Hartman, R.C. et al. (1999) ApJS 123, 79
10. Hirabayashi, H. et al. (2000) JASJ, 52, 955
11. Hoyle, F., Burbidge, G., & Sargent, W. (1966) Nature, 209, 751
12. Jauncey, D. et al. (2000) in Astrophysical Phenomena Revealed by Space VLBI, eds. H. Hirabayashi et al. (Sagamihara, Japan) pg. 147
13. Kedziora-Chudczer, L. et al. (1997) ApJ, 490, L9.
14. Kellermann, K.I., Vermeulen, R., Zensus, A., & Cohen, M. (1998) AJ, 115, 1295
15. Kellermann, K. & Pauliny-Toth, I. (1969) ApJL, 155, L71
16. Kellermann, K., Vermeulen, R., Zensus, A., Cohen, M., & West, A. (1999) New Astr. Rev. 43, 757
17. Kellermann K, Vermeulen, R., Zensus, A., & Cohen, M. (2000), in Astrophysical Phenomena Revealed by Space VLBI, eds. H. Hirabayashi, P.G. Edwards, and D.W. Murphy. Institute of Space and Astronautical Science: Sagamihara, Japan. pg. 159

18. Linfield, R. et al. (1989) ApJ, 336, 1105
19. Lister, M. & Marsher, A. (1997) ApJ, 476, 572
20. Lobanov, A. & Zensus, A. (1999) ApJ, 521, 509
21. McCray, R. (1966) Science, 154, 1323
22. Pauliny-Toth, I. & Kellermann, K. (1966) ApJ, 146, 634
23. Readhead, A. (1994) ApJ, 426, 51
24. Rees, M. (1966) Nature, 211, 468
25. Rees, M. (1967) MNRAS, 135, 345
26. Sholomitsky, G. (1965) Astron. Zh., 42, 673 (Soviet AJ, 9, 516)
27. Slysh, V. (1992) ApJ, 391, 453
28. Twiss, R.Q. (1958) Austr. J. Phys., 11, 565
29. Vermeulen, R. (1995) Proc. Nat. Acad. Sci. 92, 11385
30. Vermeulen, R. & Cohen, M. (1994) ApJ, 430, 467
31. Wagner, S. & Witzel, A. (1995) Ann. Rev. Ast. Astrophys. 33, 163
32. Walker, C. et al. (2000) ApJ, 530, 233
33. Whitney, A. et al. (1971) Science, 173, 225
34. Woltjer, L. (1966) ApJ, 146, 597
35. Zheleznyakov, V. (1966) Soviet Physics JETP, 51, 570
36. Zheleznyakov, V. et al. (2000) in Radio Astronomy at Long Wavelengths, eds. R. Stone et al., American Geophysical Union, Washington, pg. 57

The Circumnuclear Environment and Early Evolution of AGN

Lodewijk Woltjer

Observatoire de Haute Provence, F-04780 Saint-Michel l'Observatoire and
Osservatorio Astrofisico di Arcetri, Largo Enrico Fermi 5, I-50125 Firenze

Abstract. X-ray and HI absorption evidence about the absorbing matter and its distribution in Sy 2 and radio galaxies is reviewed. Column densities are found to be larger in the former. In some powerful radio galaxies the absorbing region has a diameter of several tens of pc, in many Seyferts much less than that. Evidence about low densities (~ 1 cm^{-3}) of matter at 100 pc from the centers of radio galaxies is summarized and the possible extreme youth of PKS 1718-649 is noted. The Narrow Line Region, the dilute gas into which the compact radio doubles expand and the absorbing torus all appear to be in pressure equilibrium.

1 Introduction[1]

In the canonical unified models of AGN, a nucleus emitting a hard X-ray spectrum and the associated Broad Line Region (BLR) are surrounded by a dusty "torus". Along a line of sight through the torus the nucleus and the BLR are largely unobservable – except for some scattered radiation – and only the much larger (~ 100 pc) Narrow Line Region (NLR) is seen. Such an object would be classified as a Seyfert 2 (Sy 2) or a Narrow Line Radio Galaxy (NLRG) or perhaps as a weak line radio galaxy. Along a line of sight which does not pass through the obscuring "torus" the same object would be classified as a Sy 1 or a quasar with the Sy 1.5, 1.8, 1.9 and the Broad Line Radio Galaxies transitional classes.

The absorbing matter need not be distributed strictly in the form of a torus. However, since the numbers of Sy 2 and radio galaxies are a few times larger than those of Sy 1 and quasars, the half thickness of the absorbing region must be about the same as its radius. If the torus is taken to have a plane of symmetry and if the thickness is determined by the velocity dispersion, the velocities transverse to the plane and in the plane cannot be very different.

The broad emission lines have velocity widths of several thousand km/s. Since the characteristic velocity around a nucleus with mass 10^8 M$_8$ solar masses at a distance of r_{pc} parsec is given by $v_c = 700$ $(M_8/r_{pc})^{1/2}$ km s^{-1}

---

[1] In this paper we use a Hubble constant $H_0 = 66\ 2/3$ km s^{-1} Mpc^{-1} which is in the middle of the values suggested in the literature and which gives convenient transformations to other commonly used values.

it follows that at least some of the matter in the BLR must be located well within a pc from the nucleus. The torus therefore has frequently been assumed to have a radius of a pc, though larger values would seem to be possible.

Powerful radio galaxies appear to be powered by two jets from the nucleus which interact with the matter near the end of the jets. If the radio lobes are confined by ram pressure, the very early evolution should be determined by the density at distances of a few to a few hundred pc from the nucleus. The corresponding radio galaxies are described as CSO (Compact Symmetric Objects). Recent observations of the expansion of the CSO have shown that the CSO are actually very young and that the densities must be correspondingly low (table 2).

In this paper we shall discuss the column densities and radii of the obscuring tori and find differences between the Seyferts and the radio galaxies. Subsequently we shall consider the very early evolution of radio galaxies and note a particularly interesting object which may be no more than a century old.

2 Column Densities

Column densities (N_H) may be obtained from absorption effects in X-ray spectra. Reliable data have come mainly from the ASCA and BeppoSAX X-ray satellites. Risaliti, Maiolino and Salvati (1999) studied a sample of 44 Seyfert galaxies of types Sy 1.8 – Sy 2. If we omit from their sample the radio galaxies NGC 1275 and NGC 5128 and if we retain only those objects which are classified as Sy 2 in the Véron-Cetty - Véron (VVC 2000) catalogue (i.e. S2, S1h, S1i), we are left with 27 objects with the distribution of N_H values given in table 1. As noted by Risaliti et al., Sy 1.8 and Sy 1.9 have much lower N_H values, while there appears to be no correlation between N_H and the luminosity of the Seyfert nucleus.

Table 1. The distribution of N_H values in Seyferts with an S2 spectrum in the VVC catalogue and in Radio Galaxies with S2 or S3 spectra. The Sy 2 data are from the sample of Risaliti et al. (1999), the RG from miscellaneous sources. The RG sample is more likely to be influenced by selection effects.

log N_H	< 22	22–23	23–24	> 25
Sy 2	2	4	8	13
RG	11	6	7	1

Sambruna, Eracleous and Mushotzky (1999) have analyzed the ASCA data for radio galaxies, including 19 NLRG and radio galaxies with very weak or no emission lines. To that sample we may add Hya A (Sambruna et al. 2000), 4C 12.50 (O'Dea et al. 2000), NGC 1052 (Weaver et al. 1999) and Vir A and NGC 4696 (Allen et al. 2000).

In the literature there is some confusion between Seyferts and radio galaxies. Thus, the S0 NGC 1052 is generally taken to be a radio galaxy, though its radio power at 5 GHz P_5(W Hz^{-1}) is lower than that of some Seyferts with log P_5 = 22.9. But MRK 3, an elliptical galaxy with log P_5 = 23.4, is usually included among the Seyferts. We shall take all galaxies with log $P_5 \geq 23.5$ to be radio galaxies and also those with weaker radio sources when located in ellipticals or with S 3 (LINER) spectra. As a result, we include among the radio galaxies MRK 3, MRK 273, MRK 348 and PKS 2308+05 for which Bassani et al. (1999) give N_H data.

In table 1 we present the N_H distribution for the radio galaxies with S2 or S3 spectra in the VVC catalogue and those not in the catalogue, i.e. with no or very weak emission lines. It appears that the radio galaxies have significantly lower column densities than the Seyferts. If we had not included the four objects usually listed as Sy, this would apply *a forteriori*. If we were to divide the radio galaxies according to radio power, the 17 objects with log $P_5 > 24$ would include only 3 with log $N_H > 23$, while the 8 weaker radio sources would have 5 with such column densities. While in the radio galaxy sample the selection effects discussed by Risaliti et al. may still play a role, the rarity of high column density in strong radio galaxies appears significant.

3 HI Absorption

If the torus contains atomic hydrogen, HI absorption may be observable and give both HI column densities and kinematical information. In fact, HI absorption has been detected in a dozen powerful radio galaxies (log $P_5 > 25.5$); it appears to be less frequent in weaker radio galaxies (Véron-Cetty et al. 2000). However, without high resolution VLBI data, the results are difficult to interpret, since the radio lobes or jets may dilute the continuum from the nucleus in low resolution observations and lead to an underestimate of the absorption. Also some or all of the HI absorption may occur in a galactic disk unrelated to the torus.

In Cyg A both high resolution HI and X-ray data are available. Conway and Blanco (1995) found for the HI column density log N_{HI} = 21.4 + log (T_{spin}/100), while Sambruna et al. (1999) give log N_H = 23.1 and Ueno et al. (1994) log N_H = 23.6. If the absorption takes place in an atomic torus, T_{spin} would have to be in the range of 5000 – 16000 K. Actually models of a torus heated by hard X-rays could give kinetic temperatures around 8000 K (Maloney et al. 1996). If the radius of the torus is too small, the free-free optical depth τ_{ff} may become too large to observe the nucleus and the spin temperature might be increased by radiative excitation. According to Conway and Blanco these considerations require the radius of the torus to exceed 50 pc.

In PKS 1353-341, also a powerful radio galaxy, Véron-Cetty et al. (2000) measured HI absorption and showed that radii larger than 29 pc respectively 14 pc are required depending upon the atomic or molecular character of the torus. The HI absorption profile extends over some 400 km/s in the rest frame. If in the torus the random motions are comparable to the circular velocity, as seems likely in a thick torus, then for a radius of 29 pc the mass in the galaxy within that radius would be several times 10^8 $M_\odot$.

In the less powerful radio galaxy Hya A the VLBI data of Taylor (1996) give log N_{HI} = 22.1 + log ($T_{spin}/100$), while Sambruna et al. (2000) found log N_H = 22.5. For an atomic torus this would require T_{spin} = 250 K. The projected thickness of the gas layer is some 15 – 20 pc and so the radius is likely to be of this order. If so, from the results of Maloney et al. the model parameters indicate an atomic torus, unless the gas were very inhomogeneously distributed, but the low temperature would fit better to a molecular one.

In many Seyferts the situation is very different. Gallimore et al. (1999) found no correlation between N_H and N_{HI}, the former being up to 10^3 times larger than the latter even for $T_{spin} = 10^4$ K. The most probable explanation given by Gallimore et al. is that in the torus $\tau_{ff} > 1$ so that the nuclear radio source becomes unobservable. Some gas further out, unrelated to the torus, could then be responsible for the HI absorption in front of the radio jets. Also radiative excitation may play a role by increasing T_{spin}. For the latter effect to be the explanation in MRK 348, one of the radio galaxies in their sample, the radius of the torus would have to be around 0.3 pc.

For an atomic torus we have $\tau_{ff} = 6.7\, n_6\, N_{23}$ (Véron-Cetty et al. 2000) with n_6 and N_{23} the atomic density and the column density in units of 10^6 cm^{-3} and 10^{23} cm^{-2} respectively. For MRK 348 we have N_{23} = 1.06 from the compilation of Bassani et al. (1999) and $\tau_{ff} > 1$ requires $n_6 > 0.14$. For a uniformly filled disk this corresponds to a radius of 0.25 pc, and so a rather small radius is again indicated.

The fragmentary evidence so far suggests that the most powerful AGN have tori with larger radii than Seyferts and other AGN with lower power. At the same time the latter have the higher column densities and so the torus masses might be more comparable. Such a result is perhaps not surprising, since the radiation pressure of the hard X-rays should be balanced by the pressure in the torus.

The models of the absorbing torus or disk remain rather unsatisfactory in that they treat the torus as a homogeneous medium. Actually in several Seyferts the model fits require a partial covering of the source. Also variations in the absorption over periods of a few years have been detected. An example is NGC 7582. While in 1994 the source was covered for 60% by a column density log N_H = 24.2 and for the remainder with log N_H = 23.1, by 1998 the latter column density had doubled (Turner et al. 2000). Apparently much inhomogeneity is present.

4 The Youngest Radio Galaxy: PKS 1718-649?

Compact radio doubles may be expanding into a very dense medium or they may be very young. Fanti et al. (1995) and Readhead et al. (1996) have given arguments in favor of the latter possibility and conclude that the CSO are the ancestors of the large radio doubles with the radio power slowly declining as the sources age. Recent VLBI measurements have conclusively shown that several CSO are expanding rapidly with speeds of the order of 0.1 c or more. They have kinematic ages of the order of 1000 years (table 2).

Table 2. Expansion velocity, age and density in the local medium for six CSO. Subsequent columns give the coordinates; half the separation of the components R; the mean velocity of these on the assumption of motion in the plane of the sky; the kinematical ages R/V and the atomic density in the medium surrounding the components based on the ram pressure; the reference to the data as indicated below the table.

Source	R(pc)	v/c	τ(yr)	n(cm^{-3})	ref.
0108 + 388	17	0.15	370	3	1
0710 + 439	65	0.19	1100	0.6	2
1934 + 638	63	<0.1	>2000	>0.1	3
1943 + 546	79	0.26	1020	0.3	4
2021 + 614	9	0.08	375	10	5
2352 + 495	88	0.15	1900	0.4	6

References : 1. Owsianik et al. 1998; 2. Owsianik and Conway 1998; 3. Tzioumis et al. 1999; 4. Polatidis et al. 1999; 5. Tschager et al. 1999; 6. Owsianik et al. 1999.

With ram pressure confinement of the radio lobes we have $P = \rho V^2$, with P the pressure in the lobe where the jet interacts with the surrounding gas, ρ the density of that gas and V the outward velocity of the lobes. P may be estimated from the minimum energy condition, V is measured and so ρ may be obtained. Typical values for ρ correspond to n = 0.5 - 10 atoms cm^{-3} between 10 and 100 pc from the nucleus. Readhead et al. find that V is more or less constant and that $\rho \propto R^{-1.3}$ between 100 pc and 100 kpc, where R is the distance to the nucleus. In fact, the data in table 2 suggest that this relation may continue till R around 10 pc. If this gas is the confining gas of the NLR at a temperature of 10^8 K, the density further in may become constant.

The smallest known compact double (GHz peaked) radio galaxy is PKS 1718-649 with a projected component separation of only 6.8 mas or 2.0 pc (Tingay et al. 1997). Observations by Véron-Cetty et al. (1995) showed that the source is surrounded by a massive (2.3×10^{10} M$_\odot$) rather roundish disk of HI. From the measured rotational velocities they find within 30 kpc from the center a mass to light ratio $M/L_B = 1.6 (\sin i)^{-2}$ in solar units with i the inclination of the disk. To obtain a "reasonable" M/L_B ratio, i should be less than 30°. If the compact double is oriented transverse to the disk, $i < 20°$ from the absence of a flat spectrum component in the radio spectrum. We adopt $i = 25°$ and so the separation of the two components becomes 4.8 pc. Comparison with the youngest CSO in table 2 suggests that the age may well be of the order of a century.

Tingay et al. noted that the 4.8 GHz flux of PKS 1718-649 had increased by 17% between 1972 and 1993. A similar increase has occurred at 2.7 GHz. Of course, flux changes in flat spectrum sources are common, but not in GHz peaked sources. If the source were powered by jets of constant thrust and expanding in a uniform density medium preserving equipartition conditions and if self-absorption effects at 4.8 GHz may be neglected then we would have $S_{4.8} \propto t^{2/5}$. We then find that the expansion began in 1930, that the current expansion speed of the components is 0.03 c and that the local density is 60 cm^{-3}. Unfortunately, the southern interferometric network may have difficulty measuring this expansion. However, for 2000 the prediction would be $S_{4.8} = 4.93$ Jy compared to 4.7 Jy in 1993.

The assumption of equipartition during evolution may appear to be somewhat doubtful. However, the evolutionary models of Readhead et al. give some support. A constant density of 60 cm^{-3} at 2.4 pc from the center does not seem unreasonable in view of the results in table 2. If the density were to decrease outwards at that distance the source would be even younger: A density variation as R^{-x} would imply $S_{4.8} \propto t^{(2-7x/4)/(5-x)}$. If from a few pc onwards $x = 1.3$ PKS 1718-649 will never become much stronger than it is now, unless of course the strength of the jet increases. According to Tingay et al. there is no evidence for radio emission outside the two lobes. This does not mean that the source has not been active in the past, but only that some period has elapsed since the last activity took place.

Véron-Cetty et al. have commented on the similarity of PKS 1718-649 and MRK 348; both have a moderately strong radio source and a huge HI envelope. They also share the absence of significant HI absorption; it may well be that in both cases free-free absorption hides the true nucleus. If so, the torus has to have a rather small radius not to hide also the lobes of PKS 1718-649.

A particularly clear case of a hidden nucleus and no significant HI absorption may be seen in PKS 1934-638. The observations by Tzioumis et al. (1999) show no evidence for a nucleus between the two lobes. X-ray observations with Beppo Sax have been scheduled in an attempt to detect the nucleus.

5 Conclusions

Evidence for an absorbing torus has been found in Seyferts and radio galaxies. In the Seyferts a radius of the order of a pc seems probable, while in the strong radio galaxies values of 20 pc or more are likely. If the thickness of the torus is comparable to its radius, the densities in the tori are of the order of 10^6 cm^{-3} for the Seyferts and 10^3 cm^{-3} in these radio galaxies. With such parameters the Seyferts may well have a molecular torus with a pressure nT $= (10^8 - 10^9)$ K cm^{-3} or more, while the radio galaxies would have an atomic torus with nT $= (10^7 - 10^8)$ K cm^{-3}. Such pressures correspond to those in the NLR where in typical filaments n $= 10^3 - 10^4$ cm^{-3} and T $= 10^4$ K. It has frequently been thought that the filaments in the NLR are confined by hot gas with T $= 10^8$ K. If so, this gas could also confine the torus and provide the ram pressure for the confinement of the radio lobes of the CSO (table 2). Thus, a rather coherent picture of the inner regions of AGN appears to be possible.

References

1. Allen, S.W., Di Matteo, T., Fabian, A.C. 2000, MNRAS 311, 493
2. Bassani, L., Dadina, M., Maiolino, R., Salvati, M., Risaliti, G., Della Ceca, R., Matt, G., Zamorani, G. 1999, ApJS 121, 473
3. Conway, J.E., Blanco, P.R. 1995, ApJ 449, L131
4. Fanti, C., Fanti, R., Dallacasa, D., Schilizzi, R.T., Spencer, R.E., Stanghellini, C. 1995, A&A 302, 317
5. Gallimore, J.F., Baum, S.A., O'Dea, C.P., Pedlar, A., Brinks, E. 1999, ApJ 524, 684
6. Maloney, P.R., Hollenbach, D.J., Tielens, A.G. 1996, ApJ 466, 561
7. O'Dea, C.P., de Vries, W.H., Worrall, D.M., Baum, S.A., Koekemoer, A. 2000, AJ 119, 478
8. Owsianik, I., Conway, J.E., Polatidis, A.G. 1998, A&A 336, L37
9. Owsianik, I., Conway, J.E., Polatidis, A.G. 1999, New Astronomy Reviews 43, 669
10. Owsianik, I., Conway, J.E. 1998, A&A 337, 69
11. Polatidis, A., Wilkinson, P.N., Xu, W., Readhead, A.C., Pearson, T.J., Taylor, G.B., Vermeulen, R.C. 1999, New Astronomy Reviews 43, 657
12. Readhead, A.C.S., Taylor, G.B., Pearson, T.J., Wilkinson, P.J. 1996, ApJ 460, 634
13. Risaliti, G., Maiolino, R., Salvati, M. 1999, ApJ 522, 157
14. Sambruna, R.M., Eracleous, M., Mushotzky, R.F. 1999, ApJ 526, 60
15. Sambruna, R.M., Chartas, G., Eracleous, M., Mushotzky, R.F., Nousek, J.A. 2000, ApJ 532, L91
16. Taylor, G.B. 1996, ApJ 470, 394
17. Tingay, S.J., Jauncey, D.L., Reynolds, A.K. et al. 1997, AJ 113, 2025
18. Tschager, W., Schilizzi, R.T., Snellen, I.A., de Bruyn, A.G., Miley, G.K., Röttgering, H.J., van Langenvelde, H.J., Fanti, C., Fanti, R. 1999, New Astronomy Reviews 43, 681

19. Turner, T.J., Perola, G.C., Fiore, F., Matt, G., George, I.M., Piro, L., Bassani, L. 2000, ApJ 531, 245
20. Tzioumis, A.K. et al. 1999, ASP Conf. 144, 179
21. Ueno, S., Koyama, K., Nishida, M., Yamauchi, S., Ward, M.J. 1994, ApJ 431, L1
22. Véron-Cetty, M.-P., Woltjer, L., Ekers, R.D., Staveley-Smith, L. 1995, A&A 297, L79
23. Véron-Cetty, M.-P., Woltjer, L., Ekers, R.D., Staveley-Smith, L. 2000, A&A 362, 426
24. Véron-Cetty, M.-P., Véron, P. 2000, A Catalogue of Quasars and Active Nuclei (9th Edition), ESO Scientific Report No 19
25. Weaver, K.A., Wilson, A.S., Henkel, C., Braatz, J.A. 1999, ApJ 520, 130

Advances in the Study of the IGM at High Redshifts with the VLT High-Resolution Spectrograph UVES

Sandro D'Odorico

European Southern Observatory, 85748 Garching by Munich, Germany

Abstract. UVES, the VLT echelle spectrograph, has started regular operation at the Kueyen 8.2m telescope (VLT UT2) on April 1st, 2000. The instrument offers unique performance as of efficiency and resolving power over the spectral range from the atmospheric cutoff in the UV to 1μm. More than 150 hours of scientific observations obtained during the commissioning of the instrument have been made publicly available through the VLT archive. Two examples from that data set are shown to illustrate the limiting capability of the instrument in the study of absorption lines in quasar spectra: the observations of the Ly α forest at z$\sim$2 and the detection of metal lines in one Damped Lyman α system at very high redshift (z=4.466, the highest redshift DLA for which a detailed study has ever been made).

1 The High-Resolution Spectrograph UVES at the VLT

1.1 Properties of UVES

The Ultraviolet-Visual Echelle Spectrograph or UVES is one of the two VLT instrument of the first generation built directly by ESO. The other is ISAAC. UVES is a dual beam spectrograph, mounted on an optical table at one of the Nasmyth foci of UT2. The properties of UVES as verified at the telescope are summarized in Table 1. A detailed description of the instrument and of its performance can be found in [1],[2] and [3]. By the choice of the two arm layout in the design of the spectrograph we were able to optimize the coatings, the gratings and the detector properties in the spectral range of each arm. The overall efficiency of the telescope+spectrograph (no slit and atmosphere losses) is higher than 10 % from 370 to 800nm. UVES compares well with the powerful echelle spectrograph HIRES mounted on the Keck1 telescope although the collecting area of the Keck is $\sim$ 1.5 times larger. The higher instrument efficiency compensates the difference in size of the main mirror. The UVES+UT2 photon-gathering efficiency is essentially the same that of HIRES+Keck1 at visual and red wavelengths and remarkably higher in the UV-Blue and far-red spectral regions. It is in these spectral ranges that we may expect the most interesting discoveries with because new classes of objects will become accessible at high spectral resolution with UVES. Figure 1 illustrates the importance of UVES UV observations in the studies of the

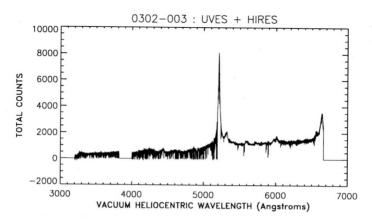

Fig. 1. The high-resolution (R=30000) spectrum of the z=3.29 QSO0302-0019. The Lyman α forest (absorptions by neutral hydrogen between the observer and the QSO) is detected at wavelengths shorter than 520 nm, corresponding to the Lyman α in emission associated to the QSO. The existing HIRES+Keck spectrum covers the spectral range 400-670 nm. The UVES data from Commissioning (λλ 310-385 nm) explore for the first time the UV region of the spectrum. This permits at the redshift of this QSO to measure for the stronger intergalactic Lyman α clouds the higher lines of the series, greatly increasing the accuracy of the H I column density measurements. Plot kindly provided by Antoinette Cowie-Songaila.

absorption spectra of QSO. Beside efficiency, other notable advantages of UVES are the possibility to carry out parallel observations in the blue and red arm by using dichroic beam splitters, a maximum resolving power of 80000 and 115000 in the blue and red arm respectively (using 0.4 arcsec and 0.3 arcsec narrow slits or image slicers) and the archiving of all data for later release to the community. Other high-resolution spectrographs to become operational at very large telescopes are HDS at SUBARU (start of operation 2001) and HROS at Gemini South (installation planned for 2002).

1.2 UVES Commissioning and Start of Operation

The first night of UVES at the telescope was September 29, 1999. The performance of the new instrument and of UT2 (also in the first period of regular operation) were remarkably smooth during the three weeks of the first commissioning period. A total of around 10 hours were lost to technical problems of telescope or of instrument. We verified that the instrument could reach the expected performance in terms of resolution, stability and efficiency. All the observations were obtained using observation blocks (a short file prepared in advance with all the information for telescope and instrument to carry out the observation) as foreseen for the regular operation of the telescope. The first period of commissioning was followed by a second shorter test in

Table 1. UVES CAPABILITIES AND MEASURED PERFORMANCE

	BlueArm	**RedArm**
Wavelength range	300-500 nm	420 -1100 nm
Echelle	41.59 g/mm, R4	31.6 g/mm, R4
Crossdispersers	CD1:1000g/mm,λ_b=430nm	CD3:600g/mm, λ_b=560nm
	CD2: 660g/mm,λ_b=460nm	CD4:312g/mm,λ_b=770nm
CCD Format	2048x3000	4096 x 4096
	windowed to 2k x 3k	2 x 1 mosaic
Scale	0.25 "/pix	0.18"/pix
Res x slit	41000	38700
Wavelength bin	0.0019 nm at 450 nm	0.0025 nm at 600 nm
Max resolution	80000	115000
	0.4" slit or IS	0.3" slit or IS
Throughput	12 % at 400nm	15 % at 600nm

December 1999. This was dedicated to the implementation of a reduction pipeline for the UVES spectra directly at the telescope. It is now possible for all standard modes of operation of the instrument to obtain within minutes from the observation the extracted, wavelength calibrated and merged echelle spectrum. This permits to assess during the night the S/N of the data and to verify the resolution of the instrument.

In February 2000 UVES was used for 8 nights for a restricted number of observing programs to verify its readiness for scientific work. The 80 hours of observations of this period (the so-called UVES science verification) are a public release of the VLT archive.

UVES started regular operation on April 1st, 2000. In the first semester (April-September 2000) UVES will be used for 70% of the available time (the remaining nights will be reserved to FORS2). The observing time is almost equally shared between visitor mode (where the P.I. is present at the telescope and can take real-time decisions on the execution of his/her program) and service mode (where all the observations have been prepared in advance by the P.I. and are executed by the observatory staff when the requested sky conditions are met). At the time of this meeting, one very successfull month of operation has been completed. Due to a unique combination of excellent weather and smooth operation, the science open shutter time in the first 25 nights of operation has reached the remarkable value of $\sim$ 75% of the total night time. A very auspicious start for the new instrument at a new telescope.

2 Scientific Results from the Commissioning Observations

The smooth behaviour of the instrument during the commissioning has resulted in a large number of scientific observations which were used to test the limiting performance of the instrument. Spectra of stars, galaxies and QSOs for a total of more than 150 hours of integration can be accessed and requested from the VLT archive through the UVES web page in the ESO site. In this paper I report on two results obtained during Commissioning which illustrate well the power of the instrument in the UV-Blue and in the far red regions of the spectrum, the spectral ranges where its performance is unique.

2.1 The Lyman α Forest at z$\sim$2

The so-called Lyman α forest in a QSO spectrum, that is the absorption by neutral hydrogen clouds (or more likely filaments of neutral gas) interposed between the Earth and the quasar, is an unique source of information on the distribution and the properties of the intergalactic medium at different epochs in the universe.

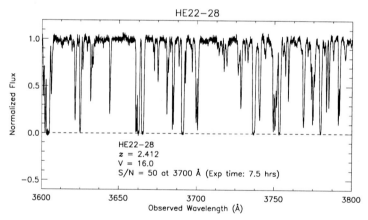

Fig. 2. The normalized spectrum of a portion of the Lyman α forest in the direction of the quasar HE 22-28 ($m_b = 17.3$). The resolution of the spectrum is 45000

The Lyman α emission of a QSO (marking is emission redshift) is shifted into the atmospheric window at redshifts larger than $\sim$1.5. Since the HIRES spectrograph at Keck has a low UV sensitivity, the statistical properties of the Lyman α forest at redshifts $\sim$2 are essentially unknown. The z$\sim$2 epoch is crucial in our understanding of the evolution of galaxies and IGM with redshift, being the link between the high redshift and the local universe.

The high UV efficiency of UVES will now permit detailed studies at high resolution of the Lyman α forest in the line of sights to a relatively large number of quasars. Already during the commissioning time, three quasars at redshift ∼ 2 have been extensively observed. An example of the reduced data is shown in Figure 2. A first detailed analysis of the properties of the Lyman α forest at z∼2 for the three line of sights has been carried out in [4]. Figure 3 from [4] shows the Log $N(z)$ versus Log (1+z) plot. The new UVES data show a behaviour consistent with the slope determined at higher redshift, indicating that Hubble expansion is the main physics dominating the IGM statistics down to z∼1.5 [4].

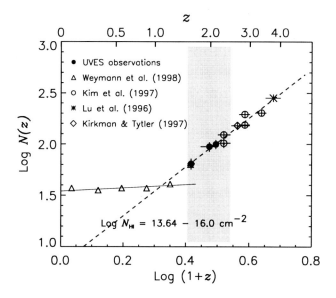

Fig. 3. The Log $N(z)$ versus Log$(1 + z)$ plot for Lyman α absorptions in different redshift bins, taken from reference [4]. Keck+HIRES observations at higher redshifts are combined with the three new UVES measurements in the range z=1.5-2.1. No significant change of slope is observed to z∼ 1.5. The triangles at lower redshifts are measurements from the HST at lower resolution.

2.2 The Studies of Damped Lyman α Systems at High Redshifts

The Damped Lyman Systems, or DLAs, are absorption systems detected in the spectra of distant qso and characterized by a large ($\geq 2 \times 10^{20}$ atoms cm^{-2}) column density of H I. DLAs are the main reservoir of neutral gas in the high redshift universe and hence one of the most important loci of star formation at early epochs. Their hydrogen and metallic, mostly low ionization absorption lines can be measured up to the redshifts of the most distant quasars at ∼5. The metal content and the relative metal abundances derived from these measurements in combination with the model predictions of stellar nucleosynthesis and of galaxy evolution, constrain the amount of star formation and the star mass function for the associated galaxy.

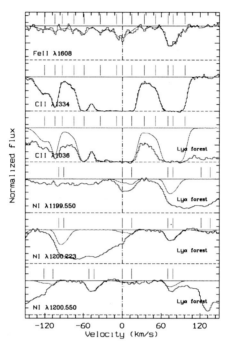

Fig. 4. Metal absorption lines associated with a DLA system at z=4.466, the highest redshift galaxy for which a detailed abundance analysis has been carried out so far. All lines are plotted in a velocity space relative to the redshift of the DLA. From UVES observations during the commissioning. The smooth lines are the the best fits used the determine the column densities. The metal absorptions in the forest are partially contaminated by Lyman α lines at different redshifts.

As a test case during UVES commissioning we have observed the quasar APM BR J0307-4945 (m_r=18.8, z=4.78) where a DLA had been discovered at z=4.466, the highest redshift known to the present time for this type of objects. The observation was particularly challenging because of the faintness of the background quasar and of the high redshift of the DLA, which shifts the most important metal lines in the far red region of the spectrum. Six exposures for a total of 21900 s at a resolution 43000-48000 were combined to obtain a spectrum of the quasar with $S/N \geq 20$. Figure 4 shows sections of the normalized spectrum including some of the key lines used in the abundance analysis. The results are discussed in detail in [5]. The data reveal a relatively complex velocity profile, spanning ∼ 240 km/sec, indicating either a large mass or strong relative motions of different clouds. The Fe $\lambda 1608$ line (observed wavelength ~879nm !) corresponds to a metallicity ~ 1/90 solar, indicating that at a look-back time of ~ 12 Gyr this absorber has already been enriched by the products of stellar nucleosynthesis. The relative abundances suggest an enrichment pattern dominated by Type II supernovae as in the early phases of the evolution of our own Galaxy.

The accurate measurements on this high redshift DLA observed toward a relative faint quasar demonstrate the unique capabilities of UVES for this type of work. We will be able to collect similar data for a large sample of these objects at large redshifts (so far only an handful at $z \geq 3.5$ are available). The abundance patterns we will derive should clarify the nature of the DLA parent galaxies (massive galaxies, prototypes of today spirals or dwarfs objects which will merge at a later time?) and their role in the metal enrichment of the early universe.

Acknowledgements

The remarkable performance of UVES at UT2 from the first night of commissioning was the result of the technical and managerial skills and of the dedication of the VLT staff and of the UVES team. It was a pleasure to work with this group of people to the successfull conclusion of the project. I like also to thank Miroslava Dessauges-Zavadsky and Tae-Sun Kim for their enthusiasm and a fine job in the analysis of the first scientific observations of UVES.

References

1. D'Odorico, S., Cristiani, S., Dekker, H., Hill, V., Kaufer, A., Kim, T. and Primas, F.(2000) Proceedings SPIE **4005**, in press
2. Dekker, H., D'Odorico, S., Kaufer, A., Delabre B. and Kotzlowski H. (2000) Proceedings SPIE **4008**, 534
3. D'Odorico, S., Kaper L., Kaufer A., (2000) UVES User Manual, VLT-MAN-ESO-13200-1825
4. Kim, T., Cristiani S., D'Odorico, S. (2000), A&A, submitted
5. Dessauges-Zavadsky, M., D'Odorico, S., McMahon, R.G., Molaro, P., Ledoux, C., Peroux, C. and Storrie-Lombardi L.J. (2000), A&A, submitted

The Evolution of Quasars and Their Relation to Galaxies

Patrick S. Osmer

Department of Astronomy, The Ohio State University, 140 W. 18th Ave., Columbus, OH 43210, USA, posmer@astronomy.ohio-state.edu

Abstract. In this article I describe some of Lo Woltjer's many accomplishments in research and in European astronomy. Then I review recent advances in the discovery of quasars and galaxies at redshifts greater than five and discuss their implications for the understanding of the formation and evolution of both galaxies and their central black holes.

1 Lo Woltjer

1.1 Seyfert Galaxies and Quasars

Among the many contributions of Lo Woltjer to astronomical research are several key papers on Seyfert galaxies and quasars, which I enjoyed reviewing in preparation for this conference. Let me mention a few which I thought were particularly interesting from our vantage point in the year 2000.

In 1959, in advance of the discovery of quasars, he [1] called attention to the unusual properties of the compact nuclei of Seyfert galaxies and noted that their lifetimes must be at least a few times 10^8 years. He argued that the nuclei were massive and that the large widths of the emission lines were consistent with the gravitational motions that would be expected from the central mass.

In 1966 Lo [2] challenged the argument of Hoyle, Burbidge, and Sargent [3] that quasars could not be at cosmological distances because the energy losses of electrons by inverse Compton radiation would be too great. He showed that if the electrons moved in narrow cone about radial magnetic field lines, the losses would be much smaller. He also noted that the radio galaxy NGC 1275, was similar to quasars except for its lower intensity and that was evidence for the two classes of objects having the same physical nature.

In 1977 Setti and Woltjer [4] wrote about the clustering of quasars and showed that the number of pairs in the Braccesi et al. [5] sample was not greater than that expected by chance, in contrast to some earlier claims for the significance of possible quasar clustering. Setti and Woltjer went on to note, however, that at magnitudes $V > 20$ quasar clustering should be expected if quasars were related to giant elliptical galaxies. Only in recent years has the evidence for quasar clustering become unambiguous, and we await results from surveys like the 2dF and Sloan Digital Sky Survey for a good determination of quasar clustering properties.

In 1979 Setti and Woltjer [6] addressed the subject of X-ray emission from quasars and the limits set by the X-ray background, which meant that the slope of the number counts of faint quasars needed to level off and that quasars probably contributed for more than half the X-ray background. Again, recent results are showing the prescience of their contribution.

Finally, let me mention the 1990 paper by Veron-Cetty and Woltjer [7] in which they studied the properties of the host galaxies of luminous quasars from ground-based data obtained with the ESO 2.2-m and 3.6-m telescopes. Their findings about the varied nature of quasar host galaxies and the range of their host luminosities, morphological types, and amount of radio emission anticipated well the later findings in higher resolution images from Hubble Space Telescope.

1.2 European Astronomy

From observing European astronomy at the distance of South and North America and being reminded by the discussion at this meeting itself, I think that the very great contributions Lo Woltjer has made to European astronomy stand out clearly. In addition to admiring his elegant appearance and his elegant approach to science, I admire his great success in getting the VLT project planned and funded. The data we are already seeing from VLT and the impressive rate at which the telescopes and instruments are being commissioned already show what a success the project is. Lo demonstrated great vision in investigating locations in Chile north of La Silla to see if they had better observing conditions and then selecting Cerro Paranal as the site for the VLT when it became clear (so to speak) that it was superior to La Silla. And, I have always been extremely impressed by his success with the 4-telescope concept of the VLT, that is, promoting it as the world's largest telescope, which it is in terms of total collecting area, while also insuring that each telescope has its own powerful instruments and can be used independently, thus providing the European astronomical community with access to 4 superb 8-m telescopes. Lo also played important roles in the success of European space astronomy. In summary, I hope that Lo and the entire European astronomical community are very pleased with their many outstanding accomplishments in recent years.

2 Quasars and Galaxies

Quasars and AGNs are rare and exotic objects that exhibit some of the most extreme physical conditions and activity known in the universe, and much research about them has focussed on their nature and properties. More recently, however, there have been many studies, both observational and theoretical, about their relation to galaxies, and we are rapidly improving our knowledge of how the formation and evolution of both galaxies and quasars/AGNs are related.

The strong evolution of the luminosity function of quasars with redshift has been clear since Schmidt's [8] 1968 paper, and there is much evidence for both optically and radio selected quasars showing a peak in space density near redshift two to three, with a marked decline toward higher redshift [9–12]. The development of effective multicolor and slitless spectrum techniques for finding high redshift quasars and their use with wide field cameras and large telescopes is producing rapid progress in the field, as is discussed below. The Hubble Space Telescope (HST) and 8–10-m ground-based telescopes provide the observational capability to reach beyond $z = 5$ and enable us to close in on the epoch when both galaxies and quasars first became visible.

On the other hand, quite different studies at low redshift are providing important connections and constraints on the late stages of quasar and galaxy evolution. For example, the recent evidence (e.g., [13]) that black holes commonly exist in the nuclei of nearby galaxies indicates that they are a normal occurrence in the formation of galaxies. How the black holes grow in mass and shine as quasars or AGNs is then connected to the physics of the accretion processes and what causes matter to fall to the nuclear regions as the parent galaxies form and evolve. Quasars and AGNs are rare because the time scale of their active state is short compared to the lifetime of their host galaxies. Furthermore, the decline of quasar activity between redshift two and the present is presumably a reflection of a decline in the accretion rate onto the central black holes.

2.1 Some Current Issues

Since much of our current observational knowledge of quasars is based on relatively small samples in which the objects are often near the faint flux limit of their surveys, there is a variety of questions which need to be answered more definitively.

For example, early results from the CADIS survey [14] found 6 quasars with $2.2 < z < 3.7$ down to $R = 22$ mag in one of their fields, which is about 6 times the number expected. Does this mean our estimates of the numbers of faint quasars is considerably in error? Or, is it a combination of a statistical fluctuation and/or possible clustering of the objects?

The ultradeep ROSAT survey [15] has the highest surface density of quasar surveys published to date, ≈ 1000 deg^{-2}. Interestingly, Miyaji et al. do not find evidence for a decrease in the space density at $z > 3$ of their X-ray selected quasars, although the significance of the finding is marginal. Whether this is a just a result of their sample size or is an indication that faint X-ray quasars behave differently from luminous optically selected quasars needs to be understood.

The recent discoveries of quasars at $z = 5.5$ [16] and $z = 5.8$ [17] are exciting results both for what they tell us about the early stages of quasars and because they serve as probes of the intergalactic medium (IGM) at redshifts previously unobserved. Although absorption in the Lyα forest is strong, the

IGM continues to be highly ionized at $z = 5.8$. These discoveries also make one wonder how or if the quasar space density is declining at the highest redshifts, in analogy with the debate on the behavior of the star formation rate in high-z galaxies. One answer to this question came after our meeting, as described next

2.2 The Sloan Digital Sky Survey

The latest results from the Sloan Digital Sky Survey (SDSS) are stunning [18,19] in terms of the numbers of high redshift quasars that have been found and analyzed systematically, 39, and of the extension of the high redshift limit for quasars to $z = 5.8$ [17]. The SDSS sample provides persuasive confirmation of the decline of the space density of luminous quasars with increasing redshift for $z > 3$. The rate of decline, a factor of 3 per unit redshift for $3.5 < z < 5.0$, is in excellent agreement with the earlier results of Schmidt, Schneider, and Gunn [10]. However, the SDSS results indicate a shallower slope for the bright end of the quasar luminosity function compared to that found for quasars with $z < 3$, which is another indication that the evolution of quasars at high redshift does not follow pure luminosity function. The slope of the luminosity function is also important for determining the contribution of quasars to the X-ray background and to the ionization of the IGM. Deeper surveys will be needed to trace out its behavior for fainter sources.

2.3 Galaxies at High Redshift

At the same time as quasar surveys have been yielding objects at $z > 5$, discoveries of high-redshift galaxies are being made at an astonishing rate. The first spectroscopically confirmed galaxy at $z > 5$ was found in 1998 [20]; it has $z = 5.34$. Subsequently, galaxies at $z = 5.6$ [21] and $z = 5.74$ [22] have been discovered, and evidence has been presented [23] for a galaxy at $z = 6.68$! For the first time since 1965, galaxies are challenging quasars for the title of the most distant objects known in the universe. This has occurred because of 1) the great light gathering power and efficient spectrographs on 8 – 10-m telescopes, 2) the utilization of HST for very deep, multicolor, imaging surveys, and 3) the development of effective techniques for selecting high-redshift objects (although Spinrad and collaborators have found some serendipitously, indicating their surface density is high). A related and important factor is that the universe forms compact galaxies with emission lines at $z > 5$. More extended galaxies of the same luminosity and redshift that only show absorption lines are still beyond the reach of slit spectroscopy However, spectra of absorption-line only objects [24] with redshifts up to $z = 4$ are now attainable.

3 Looking to the Future

As the performance of large telescopes and their instruments continues to improve, so will our knowledge of AGNs, quasars, and galaxies at high redshift. From observations of quasars and AGNs, we can expect to constrain their number density at high redshift down to L^* luminosities and below and to determine their clustering properties. Their clustering properties in combination with models of structure and galaxy formation will tell us about the masses of their parent galaxies [25,26]. Furthermore, we will be able to determine how the chemical abundances of their emission line regions evolve over the interval $0 < z < 6$.

At the same time we will be able to map the spatial association of quasars and galaxies at high and low redshifts and obtain more detailed information on quasar host galaxies at higher redshifts than has been possible to date. We will have significantly improved data on the incidence and masses of black holes in nearby galaxies and ways to relate that information to the evolution of the quasar luminosity function, thus telling us more about both the physics of accretion and the formation of black holes in galactic nuclei.

All this will come about from the completion and use of the VLT and other large telescopes, from surveys like the SDSS and 2dF, from observations with Chandra, XMM, and NGST, and from the continuing advances in theory.

I am sure Lo will be following and contributing to all these developments as he has done throughout his career.

Acknowledgements. I thank the organizers of the meeting for their hospitality and support. My research has been supported under NSF grant AST-9802658.

References

1. Woltjer, L. 1959 ApJ, **130**, 38
2. Woltjer, L. 1966, ApJ, **146**, 597
3. Hoyle, F., Burbidge, G. R., Sargent, W. L. W. 1966, Nature, **209**, 751
4. Setti, G., Woltjer, L. 1977, ApJ, **218**, L33
5. Braccesi, A., Formiggini, L., Gandolfi, E. 1970, A&A, **5**, 264
6. Setti, G., Woltjer 1979, A&A, **76**, L1
7. Veron-Cetty, M.-P., Woltjer, L. 1990, A&A, **236**, 69
8. Schmidt, M. 1968, ApJ, **151**, 393
9. Warren, S. J., Hewett, P. C., Osmer, P. S. 1994, ApJ, **421**, 412
10. Schmidt, M., Schneider, D. P., Gunn, J. E. 1995, AJ, **110**, 68
11. Kennefick, J. D., Djorgovski, S. G., de Carvalho, R. R. 1995, AJ, **110**, 2553
12. Shaver, P. A., Wall, J. V., Kellerman, K. I., Jackson, C. A., Hawkins, M. R. S. 1996, Nature, **384**, 439
13. Richstone, D. O. et al. 1998, Nature, **395**, A1
14. Wolf, C. et al. 1999, A&A, **343**, 399
15. Miyaji, T., Hasinger, G., Schmidt, M. 2000, A&A, **353**, 25
16. Stern, D. et al. 2000, ApJ, **533**, L75

17. Fan, X. et al. 2000, AJ, in press (astro-ph/0005414)
18. Fan, X. et al. 2000, AJ, submitted (astro-ph/0008122)
19. Fan, X. et al. 2000, AJ, submitted (astro-ph/0008123)
20. Dey, A., Spinrad, H., Stern, D., Graham, J. R., Chaffee, F. H. 1998, ApJ, **498**, L93
21. Weymann, R. J. et al. 1998, ApJ, **505**, L95
22. Hu, E. M., McMahon, R. G., Cowie, L. L. 1999, ApJ, **522**, L9
23. Chen, H.-W., Lanzetta, K. M, Pascarelle, S. 1999, Nature, **398**, 586
24. Cristiani, S. et al. 2000, A&A, **359**, 489
25. Haiman, Z., Hui, L. 2000, ApJ, submitted (astro-ph/0002190)
26. Martini, P., Weinberg, D. 2000, ApJ, submitted (astro-ph/0002384)

Why We Need Larger Radiotelescopes: The Atacama Large Millimeter Array

P.A. Shaver

European Southern Observatory, Karl-Schwarzschild-Str. 2,
D-85748 Garching bei München, Germany

Abstract. A major breakthrough has occurred in millimeter astronomy over the last decade, with the detection of objects at the highest redshifts in both line and continuum and the realization that much of the luminosity from the distant Universe appears at far infrared and millimeter wavelengths. Current millimeter telescopes can only just detect the most luminous sources, giving us tantalizing first glimpses of the epoch of galaxy formation. A very large millimeter telescope is now required to explore these new horizons, one with both high sensitivity and angular resolution – The Atacama Large Millimeter Array.

1 Introduction

It is appropriate that this topic is included in a meeting honouring Lo Woltjer's 70th birthday, as Lo has played a significant part in the development of European millimeter astronomy leading up to the Atacama Large Millimeter Array (ALMA). As ESO's Director General he brought ESO into the field of millimeter astronomy in 1984 with the agreement with Sweden to build and operate the Swedish-ESO (Sub)millimeter Telescope (SEST) on La Silla. In 1995 he gave the concluding remarks at a workshop on Science with Large Millimetre Telescopes, and stated "The scientific case for such an array is overwhelming. From the nearest planets to the outer reaches of the Universe it would have an enormous impact."

Until a decade ago, millimeter astronomy was essentially confined to studies of our Galaxy and its local neighbourhood. Indeed, in the early 1970s, it was not even clear whether there would be sufficient structure to detect with millimeter interferometers, as the galactic molecular clouds known at that time could be resolved out at high angular resolution.

Three related developments over the last decade have catapulted millimeter astronomy into the front ranks in the study of the distant Universe. First, CO emission was detected from the extraordinary ultraluminous source IRAS 10214+4724 at a redshift of 2.3 (Brown and Vanden Bout, 1991). The horizons of millimeter astronomy suddenly expanded to encompass the entire Universe. Second, a far-infrared and (sub)millimeter background was detected using the COBE satellite (Puget et al., 1996), indicating the presence of large numbers of luminous sources at these wavelengths. And third, the new bolometer arrays such as SCUBA on the JCMT and MAMBO on the IRAM

30m telescope began to detect the most luminous of a vast new population of (sub)millimeter sources at high redshifts (eg. Hughes et al., 1998), a population which probably accounts for most if not all of the newly discovered background.

To explore this new domain - the epoch of galaxy formation - a very large millimeter telescope is required. It will be a major new facility for all fields of astronomy, and will bring about a revolution in millimeter astronomy. It will be a millimeter equivalent of the VLT, HST and other major world facilities, with similar angular resolution and sensitivity but unhindered by dust opacity. It should be capable of seeing star-forming galaxies across the Universe, and star-formation regions across the galaxy. To achieve these objectives, an array of antennas with a large total collecting area is required. The site must be high, dry, large and flat, and a high plateau in the Atacama region of northern Chile is ideal.

2 The Atacama Large Millimeter Array (ALMA)

ALMA is a joint Europe-US project, which may soon be augmented with the addition of Japan as a third equal partner. As currently envisaged, it will be comprised of 64 12m antennas, giving a total collecting area of 7000 m^2. The antennas surfaces will be accurate to 25μ rms or better, to provide high sensitivity at the highest frequencies. The antennas will be movable, so that the configuration can be changed and provide a "zoom lens" capability. The most compact array will be concentrated within 150m, and the largest array will have baselines up to 10-20 km, giving an angular resolution of 10 milli-arcsec at the highest frequencies. The receivers will cover all atmospheric bands in the frequency range 30-900 GHz. The site chosen is the Llano de Chajnantor, near San Pedro de Atacama in northern Chile. At an altitude of 5000m it has exceptionally low atmospheric opacity, and is certainly one of the very best sites in the world for such an array. The combination of large collecting area, high antenna precision, low receiver noise and low atmospheric opacity results in an rms sensitivity of a few micro-Jy in 10 hours of integration at 230 GHz.

More details of the ALMA project are given below. In the next section an overview is given of the range of science that will be possible with ALMA.

3 Science with ALMA

The main science drivers for ALMA are the origins of galaxies, stars and planets: the epoch of first galaxy formation and the evolution of galaxies at later stages including the dust-obscured star-forming galaxies that HST and VLT cannot see, and all phases of star formation hidden away in dusty molecular clouds. But ALMA will go far beyond these main science drivers

– it will have a major impact on virtually all areas of astronomy. It may well have as big a user community as the VLT itself.

Some idea of the range of science that can be done with ALMA is given below, with emphasis on the high-redshift Universe, quasars and AGN. Much more detail can be found in the proceedings of the 1995 ESO-IRAM-NFRA-Onsala workshop on Science with Large Millimetre Arrays (Shaver, 1996) and the proceedings of the 1999 meeting on Science with the Atacama Large Millimeter Array held in Washington (Wootten, 2000). Other references can be found at the ALMA websites http://www.eso.org/projects/alma/ and http://www.alma.nrao.edu.

3.1 The High-Redshift Universe

The early Universe is particularly accessible to observations in the millimeter and submillimeter wavebands. Whereas the broadband flux from distant galaxies is diminished in the UV and optical both due to the redshift and obscuration by internal dust, the same dust produces a large peak in the rest-frame far-infrared, which, when redshifted, greatly enhances the millimeter and submillimeter emission from these objects. This negative K-correction is so strong that galaxies at $z \sim 5-10$ can appear brighter than those at $z \sim 1$. Thus, ALMA may provide one of the best ways to find the first galaxies after the "dark ages" at $z > 5$, and certainly it will be highly complementary in finding dust-obscured star-forming galaxies that would be hard to detect with HST or the VLT.

It is now known that the extragalactic background radiation in the far-infrared and submillimeter wavebands is comparable to that in the optical/near-infrared. At low redshifts the optical/near-infrared dominates, so at high redshifts the far-infrared/submillimeter must be dominant. Indeed, this is consistent with the discovery of a population of very luminous dust-enshrouded star-forming galaxies at $z > 1$ using the SCUBA submm-wave camera on the JCMT at a wavelength of 850μ, and that fact that these sources already appear to account for most of the submillimeter background. Observations of both the background and these SCUBA galaxies indicate that interstellar dust absorbs most of the optical/UV photons emitted in the high-redshift Universe. The absorbed energy heats the dust, so it appears as thermal radiation in the rest-frame far-infrared and is redshifted into the submillimeter. Hence, much of the energy produced in the high-redshift Universe may be detectable only at wavelengths accessible to ALMA.

The present highly successful millimeter and submillimeter surveys have just found the tip of the iceberg. They are sensitive only to the rare sources brighter than about 1 mJy, whereas ALMA will be able to reach 10 μJy – one hundred times deeper. Of even greater importance is the angular resolution. The SCUBA observations are made with a 15-arcsec beam, and it has been extremely difficult to identify the galaxies and make follow-up observations in the optical. Various alternative approaches to identification using intermediate observational steps have been taken, *e.g.* interferometer measurements at millimeter wavelengths with the IRAM Plateau de Bure array, and VLA observations at centimeter wavelengths, an approach which relies on the radio-far-infrared correlation. These are successful in some cases, but overall the identification of the SCUBA sources remains a major obstacle. A related problem is source confusion. A significant increase in sensitivity must be accompanied by a corresponding increase in angular resolution, and ultimately the angular resolution provided by ALMA is required.

Deep Surveys ALMA will be the only telescope with both the sensitivity and angular resolution to carry out deep surveys in the millimeter and submillimeter wavebands below 100 μJy – flux densities at which L^* galaxies would be detectable to moderate redshifts. Because of the negative K-correction its sensitivity to sources at $z \sim 1 - 10$ will be largely independent of redshift, facilitating the interpretation of source counts. The surface density may approach that of the Hubble Deep Field at the survey limit. Such deep surveys will discriminate between different models of galaxy evolution, determine the epoch of galaxy formation and study large-scale structure at high redshifts.

ALMA will also be essential in studying sources detected with other instruments. The optically-selected Lyman-break galaxies are expected to have (sub)millimeter flux densities that will be detectable in minutes using ALMA. It is important to know what fraction of the bolometric luminosity is being missed in the optical/UV surveys and re-radiated in the far-infrared/submillimeter bands. Infrared-selected galaxies found with 175 μm flux densities of $>$ 100 mJy using ISO are expected to have 850 μm flux densities of several mJy if they are at $z \sim 1$. Much larger samples of similar objects will come in the next few years from facilities such as SIRTF. ALMA will be sufficiently sensitive to rapidly search the error circles and identify the sources. Large samples of submillimeter-selected sources will come from BOLOCAM on the Large Millimeter Telescope and from the FIRST and Planck space missions. While the Planck survey is primarily intended to study the microwave background, it will also detect tens of thousands of galaxies at 350 GHz with a flux density of greater than 100 mJy. ALMA will be able to identify such sources with integration times of seconds, and measure their spectral distributions. This will produce a catalogue of all the most luminous submillimeter galaxies and AGN over the southern sky. ALMA, FIRST and Planck will be highly complementary facilities.

In many cases it should be possible to determine the redshift purely from the millimeter and submillimeter observations, without recourse to the optical/infrared. The "ladder" of molecular transitions makes it likely that a redshifted line will appear in one of the observing bands. The detection of two will determine the redshift. Already CO lines have been detected from some of the highest-redshift sources known (eg. Omont et al., 1996), and with ALMA this will become routine.

Gravitational Lensing Gravitational lenses may be much more numerous in the submillimeter than in the optical or radio wavebands because of the very steep source count. Lensing by clusters of galaxies, particularly in the high-magnification regions around the critical lines, may make it possible to detect galaxies with unmagnified flux densities less than a micro-Jansky, corresponding to luminosities far below L^* at high redshifts. This may provide one of the best opportunities to find (mini)galaxies or (mini)quasars in the reionization epoch. It may be possible to resolve the emission from dust and gas in the star-forming fragments of primordial galaxies, allowing their morphologies and internal dynamics to be reconstructed. Searches for lensed background galaxies around known elliptical and spiral galaxies will yield counts of faint dusty galaxies and probe the dark-matter halos. In the case of strong-lensing events, ALMA should facilitate the detection and study of the lensing galaxies. The large number of baselines and high sensitivity and resolution of ALMA will make it possible to monitor known gravitational lenses with snapshot observations of high fidelity.

Quasar Absorption Lines ALMA will observe molecular absorption lines in the spectra of many quasars. This is a new field with great potential, which was pioneered at the SEST. Over 30 molecular transitions have already been detected in individual absorption systems. One of them, at $z = 0.9$, was discovered purely by spectral scans in the millimeter wavebands, in a radio Einstein ring which had not yet been optically identified. In such systems one can study detailed chemistry at cosmological distances, isotopic ratios and the microwave background temperature *vs.* redshift, and gravitational lens time delays. At present only a few such systems are accessible, but the high sensitivity of ALMA will make it possible to study hundreds of such systems. Because this is an absorption measurement there is no beam dilution, so such systems can be detected out to the highest redshifts at which luminous quasars exist.

3.2 Quasars and AGN

Like star-forming galaxies, quasars and radio galaxies also exhibit the prominent redshifted far-infrared emission peak from heated dust, and the negative K-correction makes them detectable out to the highest redshifts. Bolometer arrays on existing millimeter-wave telescopes have detected quasars up to $z = 5.5$, and CO line emission has been detected at $z = 4.7$. Such observations have an important bearing on the relationships between star formation, merging activity and the quasar phenomenon. With the high angular resolution of ALMA it will be possible to examine these processes in detail, and determine whether mergers play as an important a role at high redshifts as they do at low redshifts. Millimeter-wave observations of quasars may have an important advantage in studying the quasar host galaxies. Whereas at optical and centimeter wavelengths the bright nuclear emission can cause severe dynamic range problems, at millimeter wavelengths the contrast is less and it should be easier to image and study both the continuum and line properties of the host galaxies.

Quasars and AGN can also be studied "in depth", because of the low synchrotron and dust opacity and the unprecedented angular resolution of millimeter VLBI – tens of microarcsec, corresponding to spatial scales of 10-10^4 Schwarzschild radii, the size of the expected accretion disk. ALMA would provide milli-Jansky VLBI sensitivity, corresponding to brightness temperatures as low as $10^2 - 10^4$ K, making the environs of AGN and extragalactic mega-masers accessible.

In more nearby AGN, the optically-obscured molecular tori and rings, and the circumnuclear starburst activity, can be studied in detail. The circumnuclear starbursts in most of the known Seyferts will be resolved – NGC 1068 will be observed with parsec resolution. The presence of central black holes can be studied kinematically in a large number of galaxies. Centaurus A will be a wonderful target for ALMA, providing a unique laboratory to study molecular gas in emission and absorption in a nearby edge-on active nucleus. All of these studies will have a direct bearing on unified schemes for AGN, fueling mechanisms, and the relation between AGN and circumnuclear starbursts.

3.3 The Nearby Universe

ALMA will provide the same detail in galaxies out to $z \sim 1$ as is presently possible in nearby galaxies, providing direct and detailed information on galaxy evolution. Many questions can be addressed: the origin, distribution, and kinematics of molecular gas in early-type galaxies; the compact molecular cores and infall mechanism; the molecular content of dwarf ellipticals; intergalactic molecular clouds in groups of interacting galaxies, etc. Major dynamical features - spiral arms, bars and rings - will be resolved with enough resolution (parsec-scale at the distance of M51) and sensitivity to constrain theoretical scenarios of galaxy evolution. The mass spectrum of molecular clouds in nearby galaxies will be determined, and the H_2/CO ratio constrained. Radio supernovae are first seen at high frequencies, and ALMA will detect them out to large distances; VLBI may be used for absolute distance determination.

In the Magellanic Clouds, large statistical samples of many types of objects at essentially the same distance can be studied and compared in detail with the corresponding objects in the Galaxy: star-forming regions, SiO masers, circumstellar shells, supernova remnants. ALMA will make possible a comparative study of dense molecular clouds in the low-metallicity, high-radiation density environment of the Magellanic Clouds with galactic molecular clouds at the same linear resolution. SN 1987A in the Large Magellanic Cloud will be a prime target for ALMA. Evidence of a central source may require millimeter-wave observations because of possible free-free and dust absorption. The shock wave is predicted to hit the [OIII] ring sometime in the middle of this decade, providing an incentive for early operation of at least part of the array!

3.4 Galactic Studies

The nearest galactic nucleus, SgrA* and its environs, can be observed free of obscuration. High angular resolution is essential to distinguish this object from confusion in this crowded region. Dynamical studies, based both on radial velocities and perhaps proper motions, will help to further constrain the distribution of mass at the Galactic centre.

Interstellar molecular absorption lines will be studied along a great many sight-lines towards extragalactic sources. An interferometer is essential to resolve out contaminating extended emission. Unlike absorption-line studies towards galactic stars, the sight-lines are guaranteed to sample random clouds unperturbed by the background sources, which in this case are extragalactic. Such observations will be ideal for unbiased statistical studies. ALMA will be used to study galactic molecular clouds, astrochemistry, the conditions at the start of cloud collapse near star-formation regions and the interactions of newly-born stars with nearby molecular clouds. It will allow vastly improved chemical abundance analysis, limited only by intrinsic line blending. Observations of Galactic and extragalactic molecular clouds will provide comparative studies of chemistry and abundance variations.

The elusive protostars may best be found by virtue of their cold dust emission at millimeter wavelengths. High angular resolution is required to distinguish objects at different evolutionary phases within the same star-forming region: the collapsing cloud cores, cool dust envelopes, hot dust cocoons, hot molecular cores, bipolar flows, ultracompact HII regions, etc. ALMA will reveal the dynamics of of the dust-obscured protostellar accretion disks, the rate of accretion and infall from the molecular clouds, and the mass distribution over the disk.

An understanding of star formation requires understanding infall and outflow phenomena in a unified fashion. The molecular outflows will be observed in unprecedented detail. High resolution is necessary to identify where the jet is launched and also to study the physical and chemical structure of this inner region. Maser emission from millimetre recombination lines may be particularly useful in studying the inner ionized disk and dense outflow - regions that are obscured by dust in the optical and by free-free absorption in the radio. Detailed studies of such nearby objects will give important clues concerning the physical processes involved in the much more distant quasars and AGN.

The processes of planet formation may be revealed by ALMA observations of dusty circumstellar disks. Simulations have shown that it should be possible to detect the gaps cleared by large bodies condensing around the stars, and perhaps to detect the protoplanets themselves. High angular resolution and good image quality are also essential to distinguish the circumstellar structures from more extended molecular cloud material.

ALMA will detect tens of thousands of stars over the entire H-R diagram, and provide major advances in virtually all areas of stellar astronomy. Circumstellar shells around evolved stars, observed at millimeter wavelengths, provide a unique probe of time-dependent chemistry. They give the mass loss rate, chemical and isotopic composition, and the formation and destruction of molecules. ALMA will resolve thousands of such shells across the Galaxy, so that dependence on stellar type, local environment and galactocentric distance can be studied. The dust and many molecular species in planetary nebulae will be observed by ALMA with an angular resolution exceeding that of the current HST images.

ALMA will also have a very important role to play in studies of the solar system. Combined observations using ALMA in conjunction with spacecraft will greatly enhance studies of the planets and their satellites. Ground-based work has the advantages of regularly updated instrumentation and the possibility of monitoring and follow-up observations. Millimeter continuum observations probe the deep atmospheres of the giant planets or the surface layers of the terrestrial planets. Millimeter heterodyne spectroscopy allows the detection of narrow spectral lines and measurement of small molecular abundances. Temporal monitoring reveals composition changes as a function of season and climate. ALMA will also detect and resolve the atmospheres of planetary satellites, including features such as the SO_2 plumes from volcanic activity on Io. Millimeter observations of the sun probe the upper photosphere and low chromosphere, and measure the temperature of the ambient electrons directly. They provide a powerful diagnostic tool for high energy electrons in solar flares.

ALMA will study the size and albedo of asteroids, and probe their subsurface temperatures. Continuum observations of comets will explore dust particle sizes not accessible to optical or centimeter-wavelength observations. Spectroscopic observations will determine the molecular composition of the nuclear ices sublimating into the coma, and search for new molecular species. High resolution, high sensitivity and rapid imaging (snapshot mode) are essential. Comparisons with molecular cloud cores will provide clues to the origin of comets.

Finally, ALMA may play an important role in the search for extrasolar planets through accurate astrometry, perhaps even direct detection.

4 The ALMA Project

ALMA is a merger of the major millimeter arrays projects – the European Large Southern Array (LSA) and the U.S. Millimeter Array (MMA), and possibly the Japanese Large Millimeter and Submillimeter Array (LMSA), into one global project. This will be the largest ground-based astronomy project of the decade after VLT/VLTI, and, together with the Next Generation

Space Telescope (NGST), one of the two major new new facilities for world astronomy coming into operation by the end of the decade.

The European participation in the project involves a consortium of research organisations: the European Southern Observatory (ESO), the Centre National de la Recherche Scientifique (CNRS, France), Max-Planck- Gesellschaft (MPG, Germany), the Netherlands Foundation for Research in Astronomy/Nederlandse Onderzoekschool Voor Astronomie (NFRA/NOVA), the U.K. Particle Physics and Astronomy Research Council (PPARC), the Swedish Natural Science Research Council (NFR), and the Spanish Instituto Geografico Nacional (IGN) and Oficina de Ciencia y Tecnologia (OCYT).

The National Radio Astronomy Observatory (NRAO) leads the North American side of the partnership. NRAO is operated by Associated Universities, Inc. (AUI) under a cooperative agreement with the U.S. National Science foundation. The Owens Valley Radio Observatory (OVRO) and the Berkeley-Illinois-Maryland Association (BIMA) participate in the project through the Millimeter Array Development Consortium. Negotiations are currently underway to add Canada to the North American side of the partnership. Chile participates in the project as host state.

The project is currently in the Design and Development Phase, which extends from 1999 through 2001. The objective is to examine critical technical areas, and to completely define a joint program for construction and operation. Products of this phase will include top-level science requirements, preliminary designs tested on prototype components or subsystems including prototype antennas provided by Europe and the U.S., the organizational and management structure, and the schedule and cost derived from a detailed project work breakdown structure. The tentative cost of the joint Europe-U.S. project is $552 million, including contingencies. Japanese participation would provide an enhancement of the project, which may include more 12m antennas and a sub-array of smaller (6-8m) antennas for short spacings and very high frequencies.

The technical requirements for ALMA are within reach, although it does pose challenges in several areas: the use of composite materials for accurate reflector antennas; the development of accurate metrology systems; highly sensitive SIS receivers for submillimeter wavelengths; photonic signal transportation and manipulation; very high speed digital electronics; and software systems for remote control of the telescope, data handling and analysis. The biggest challenge is to turn these laboratory developments into products which can be mass produced cheaply and with high reliability. The two prototype antennas must meet identical specifications, but are of inherently different designs. They will undergo rigorous tests at the VLA site in New Mexico. This process will ensure that the best possible technologies are incorporated into the final production antennas. A Japanese LMSA prototype antenna is also being built, and will be moved to the ALMA site in Chile in 2001.

A variety of tests continue at the Chajnantor site, including measurements of the atmospheric opacity and phase stability, geotechnical and environmental studies. In view of the exceptional quality of the site, the Chilean government has set aside a large part of the Zona de Chajnantor as a scientific preserve under the stewardship of CONICYT. Agreements with the Chilean government will secure the long-term use of this land for ALMA and lead to the establishment of the ALMA Observatory.

The construction phase of ALMA is scheduled to begin in 2002. The first production antennas should arrive on the site in 2004, and limited scientific operations will be possible by 2006. The complete array should be available in 2010 – a powerful new addition to the world's astronomical facilities.

5 Summary

There have been dramatic advances in millimeter astronomy over the past decade, with the detection of objects at the highest redshifts in both line and continuum and the realization that much of the luminosity from the distant Universe appears at far infrared and millimeter wavelengths.A major new millimeter-wave telescope is now required to explore this new frontier: the Atacama Large Millimeter Array (ALMA). This new facility, a global collaboration, will bring major advances to virtually all areas of astronomy.

References

1. Brown, R.L., Vanden Bout, P.A. (1991), AJ 102, 1956
2. Hughes, D.H. et al. (1998), Nature 394, 241
3. Omont, A. et al. (1996) Nature 382, 428
4. Puget, J.-L. et al. (1996) A&A 308, L5
5. Shaver, P.A. (ed.) (1996) Science with Large Millimetre Arrays (ESO Astrophysics Symposia, Springer, Berlin)
6. Wootten, A. (ed.) (2000) Science with the Large Atacama Array (Publ. Astron. Soc. Pacific)

AGN's Central Machine Physics from Space VLBI

Nikolay S. Kardashev

Astro Space Center of Lebedev Physical Institute, Moscow, Russia

Abstract. The magnetosphere of a supermassive black hole is the region in which a super powerful particle accelerator operates. The scale of this region is only about a few Schwarzschild radii. A synchrotron emission of this volume is probably generated by relativistic protons. The most convenient objects for the investigation of these processes are interday variable radio sources. A short description and the operation of the RADIOASTRON Earth-Space interferometer are also discussed.

1 Introduction

Knowing the physical processes and structure of the regions around the supermassive black holes (SMBH) in the galactic centers is one of the most interesting tasks of astrophysics. These regions are extremely complicated objects, which include a very strong gravitation field of the single (or binary) spinning SMBH with ergosphere, a strong magnetic and probably electric fields, an accretion disc (containing plasma, neutral gas, dust and stars), a central star cluster (all types of stars, including neutron stars and solar mass black holes), and probably cold dark matter. The main processes in these regions are absorption of accreting matter by SMBH in the equatorial plane and particle acceleration up to relativistic energy and then ejection near the spinning axis. Observational data and theoretical models demonstrate that such a central machine should operate as a high efficiency particle accelerator or cosmic supercollider (Blanford & Znajek 1977, Macdonald & Thorne 1982, Kardashev 1995). The correlation of luminosity in the gamma and radio spectral bands, superluminal proper motions and intraday variability (IDV) of BL Lac type radio sources support a unification model which proposes an orientation of the SMBH spinning axis and relativistic particle ejection of these objects near to the direction of the observer. A gamma radiation corresponds to Compton scattering processes and the radio one corresponds to synchrotron emission by the same relativistic particles. The time scale variations shortening in the observer frame is dependent on the Doppler boosting effect, and the Compton energy losses of radiating relativistic particles are dependent on a solid angle of electromagnetic radiation in the source (Woltjer 1966). Both effects are very essential for the real model of an AGN central machine. The most principal unresolved problem is a determination of the value and configuration of the magnetic and also probably electric fields in the nearest environments of SMBH. From simple estimations a magnetic flux in this region is accumulated during the entire cosmological history of the host galaxy without any losses and therefore the magnetic field strength can approach the limiting value $H_{max} \sim c^4 G^{-1.5} M^{-1} \sim 2 \cdot 10^{10} M_9^{-1}$ G ($M_9 = M/10^9 M_\odot$). In this model an induction electric field can accelerate particles with electric charge eZ up to energy $E_{max} \sim ec^2 G^{-0.5} Z \sim 10^{18} Z$ GeV, which is near to the Planck energy $E_{pl} = (\hbar c/G)^{0.5} c^2 \sim 10^{19}$ GeV (Kardashev 1995). The presence of a regular magnetic field with permanent configuration was really recently discovered by VLBI observations (Taylor 1998, Homan & Wardle 1999), but the angular resolutions of these experiments are insufficient for the determination of the most central structure. It seems that this task can be resolved only by Space VLBI (SVLBI), while taking advantage of its high angular resolution together with data from other spectral bands.

2 Object Selection for Investigations with SVLBI

Two methods of revealing ultra compact objects have been used:

– a search for unresolved nuclei components of extragalactic radio sources by ground VLBI observations (Moellenbrock et al. 1996, Kellermann et al. 1998) and by space-ground radio interferometer with VSOP satellite (Hirabayashi et al. 2000, Preston at al. 2000),

– a selection of the extragalactic objects with strong intraday radio flux variability – IDV objects (Wagner & Witzel 1995, Romero et al. 1997, Kedziora-Chudczer et al. 1997, Kraus et al. 1999, Dennett-Thorpe & de Bruyn 2000, Kedziora-Chudczer et al. 2000, Qian et al. 2000).

The first method's result is that a large part of AGN's has unresolved nuclei components up to baselines 26,000 km. The second method permits to select some more compact objects. A mechanism of interday variability can demonstrate either intrinsic changing (of emission, absorption or scattering) in the source, or gravitational microlensing, or effects of scintillations triggered by the interstellar plasma clouds. Apparently all these phenomena were already detected. A list of the dozen most strong IDV objects is collected in Table 1: source name, equatorial and galactic coordinates, V magnitude and redshift.

The central idea for the search of the central machine by space VLBI is a proposal that the IDV radio sources are the most convenient objects, because their orientation with respect to the observer and intrinsic transparency for this direction permit probably to search very deep central regions. Let us accept that for the investigation of the structure and physical processes of the central machine, an angular resolution corresponding to one tenth of the SMBH diameter is needed. For $M = 3 \cdot 10^9 M_\odot, d = 4Gm/c^2 \sim 2 \cdot 10^{15}$ cm and at the distance $D = 30 Mpc \sim 10^{26}$ cm, it is essential that the angular resolution is $\varphi \sim 0.1 d/D \sim 0.4$ μas. For the relativistically expanding object with Lorentz factor $\gamma = 10$, and at the expanding time $\Delta t = 1$ day, the

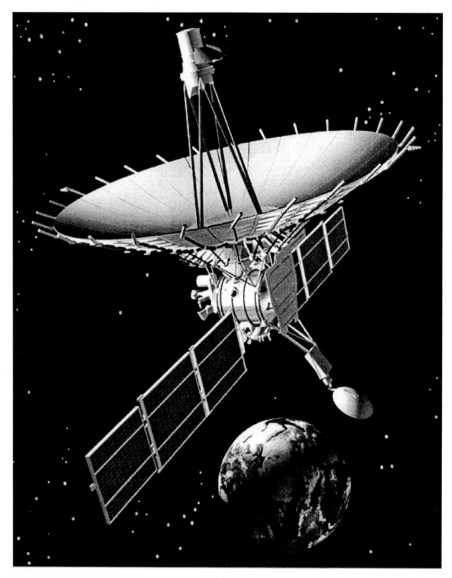

Fig. 1. RADIOASTRON satellite

Table 1. The strong IDV objects

Source	RA (2000.0) (h m s)	Dec (d m s)	L (d)	B (d)	V (m)	z
[HB89]0235+164	02 38 38.9	+16 36 59	157	−39	19.0	0.940
[HB89]0405−385	04 06 59.0	−38 26 28	241	−48	18.0	1.285
[HB89]0537−441	05 38 50.3	−44 05 09	250	−31	15.5	0.894
[HB89]0716+714	07 21 53.4	+71 20 36	144	+28	15.5	—
[HB89]0804+499	08 08 39.6	+49 50 37	169	+33	17.5	1.430
TXS 0917+624	09 21 36.2	+62 15 52	152	+41	19.5	1.446
[HB89]0954+658	09 58 47.2	+65 33 55	146	+43	16.7	0.368
[ME94]1308+328	13 10 59.4	+32 33 34	86	+83	19.2	1.650
PKS 1519−273	15 22 37.7	−27 30 11	340	+24	18.5	—
[HB89]1633+382	16 35 15.5	+38 08 04	61	+42	18	1.814
J 1819+3845	18 19 26.6	+38 45 02	66	+22	21	0.540
[HB89]2007+777	20 05 30.9	+77 52 43	110	+23	16.5	0.342

angular diameter in the observers frame $\varphi = 2\gamma c \Delta t/D \sim 100\ \mu as$ at $D = 30$ Mpc, and $\varphi \sim 3\ \mu as$ at $D = 1$ Gpc ($z = 0.2$ at $H_0 = 65$ km/s Mpc), and $\varphi \sim 0.1\ \mu as$ for the same distance but for $\Delta t = 1$ hour.

The brightness temperature in the source frame and at the Gauss profile of the intensity distribution is

$$T = \frac{2ln2}{\pi k}\frac{\Delta F \lambda^2 (1+z)}{\varphi^2} = 1.36 \cdot 10^{15}\frac{\Delta F \lambda^2 (1+z)}{\varphi^2}\ (K).$$

In the last relation the observing flux variation is ΔF (in Jy), the wavelength of observations is λ (in cm), the angular diameter on a half intensity level is φ (in microarcseconds). For $\varphi = 0.1 - 100\ \mu as$, $\lambda = 1$ cm, $\Delta F = 0.3$ Jy and $z \ll 1$ the brightness temperature is $T = 4 \cdot 10^{10} - 4 \cdot 10^{16}$ K.

The brightness temperature lower limit from IDV observations may, without Doppler boosting, be as high as $10^{19} - 10^{21}$ K (Wagner & Witzel 1995). Apparently an IDV of such a type really exists because for some sources, [HB89] 0716+714 for example, the observed correlation between the variability in optical and radio bands (Quirrenbach et al. 1991, Wagner et al. 1996) and a time scale of the IDV is very near to the expected Schwarzschild radius light time.

From the other side it was demonstrated that variations of some IDV sources are caused by interstellar scintillations. The source [HB89] 0405 -385 was observed simultaneously with the VLA in New Mexico and with ATCA in Australia at 6 cm wavelength and the time delay was revealed from the cross-correlation of flux variations which is in accordance with the expected motion of the observers with respect to the interstellar plasma clouds and the displacement of these two places of observations (Jauncey et al. 2000). At last the radio-microlensing effect and even with caustic crossing was also probably observed. For the sources [HB89] 0537 -441 and B 1600+434 a sporadic flux increasing and following drop to the primary level (Romero et al. 1995, Koopmans et al. 2000) was detected.

It is essential that all these three effects be displayed simultaneously for the point sources, i.e. that each effect demonstrates a very small size of the source (less than 10 μas).

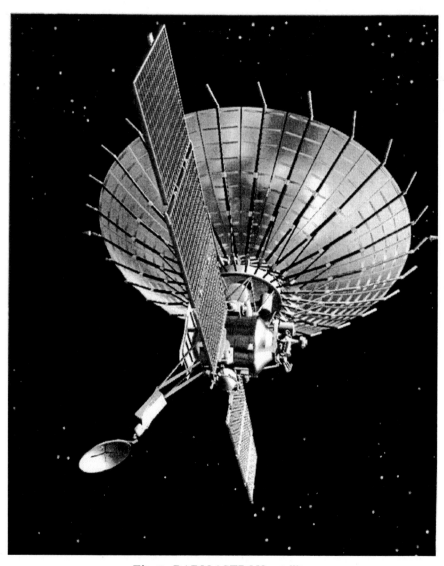

Fig. 2. RADIOASTRON satellite

3 On Proton-Synchrotron Emission from IDV Objects

IDV radio sources, independently of their variability, indicate very high brightness temperatures which are much higher than a Compton limit 10^{12} K (Kellermann & Pauliny-Toth 1969). The usual explanation of this as a result of relativistic beaming only (Woltjer 1966, Rees 1967) requires a too high Doppler factor, which is not confirmed from proper motion observations. The second explanation is a coherent electron-synchrotron emission similar to the radio emission in pulsars and is also possible for strong amplification of intensity. However for the expectable conditions near the SMBH (large value of magnetic field and very high density of electromagnetic radiation) the real life time of relativistic electrons is very short. In this connection the region of SMBH environments is very convenient for the acceleration of cosmic rays, and very high brightness temperatures of radio emission can be explained by proton-synchrotron radiation (Jukes 1967, Rees 1967, Pacini & Rees 1970, Kardashev 2000). The self Compton brightness temperature limit is

$$T = a(n)(mc^2/k)(m^2c^4e^{-3}H_\perp^{-1})^{1/7},$$

where n is the power index of the relativistic particle number density spectrum $N = KE^{-n}$; for $n = 1, 2$ and 3 a value $a(n)$ is accordingly $0.85, 0.28$ and 0.19, $H_\perp$ is a magnetic field component orthogonal to the particle velocity. For the monoenergetic particle spectrum the value is 0.725. For the protons the temperature limit is $T_p = a(n) 1.7 \cdot 10^{16} H_\perp^{-1/7}$ K for the power spectrum and $T_p = 1.2 \cdot 10^{16} H_\perp^{-1/7}$ K for the monoenergetic one.

For the ejected particles in the magnetosphere of a SMBH, and assuming the conservation of adiabatic invariant, the particle velocity vectors approach the direction of the magnetic field lines. The main synchrotron radiation for such particles will be connected with the curvature of the magnetic field lines (a curvature radiation). For such emission and for a monoenergetic particle spectrum a limiting temperature is

$$T = a^{7/8}(mc^2/k)(0.75mc^2e^{-2}\rho_c)^{1/8}$$

For a SMBH mass $M = 3 \cdot 10^9 M_\odot$, dipole magnetic field with a curvature radius $\rho_c = 4R^2/(3r)$, a distance from the center $R = 10R_g = r = 10^{16}$ cm, here r is the distance from the dipole axis, the brightness temperatures for the electron/positron ($e^\pm$) and proton emissions are $T_e = 1.6 \cdot 10^{13}$ K and $T_p = 8 \cdot 10^{16}$ K respectively. The characteristic frequency of the curvature radiation is $\nu_c = (3c/4\pi)(E/mc^2)^3/\rho_c$ and for $\nu_c = 100$ GHz, the Lorentz factor is $E/mc^2 = 6 \cdot 10^5$ and a proton energy is $E = 5 \cdot 10^{14}$ eV.

The main advantage of the proton-synchrotron model is the possibility of its realization in the conditions near the SMBH. For the relativistic protons the synchrotron energy loss is $-\dot{E} \propto H_\perp^2 E^2 m^{-4}$, that is 10^{13} times smaller than for $e^\pm$. The same estimation is correct also for the Compton losses. A characteristic frequency of synchrotron emission is $\nu_c \propto H_\perp E^2 m^{-3}$, i.e. 10^{10} times lower for protons than for $e^\pm$. The ratio of the energy loss times for the protons and $e^\pm$ at the same ν_c and $H_\perp$ is $\tau_p/\tau_e = (m_p/m_e)^{2.5} = 10^8$. This is a significant factor which permits to expect a dominant proton-synchrotron radiation from the internal volume near the poles of the spinning SMBH, and observations with ultra high angular resolution enable to investigate the region of cosmic ray acceleration.

4 The RADIOASTRON Mission as a Ground-Space Radio Interferometer with Ultra High Angular Resolution

The investigation of the central machine in galactic nuclei is one of the most interesting tasks for space VLBI with the largest baselines: it depends on the high values of the brightness temperatures of emission which permit to obtain data even with ten-meter size space radiotelescopes. The sensitivity of the ground-space interferometer is inversely proportional to the diameters of the space and ground-based antennas. I.e. a 100 m mirror on the ground and a 10 m mirror in space are equivalent in sensitivity to two 32 m mirrors. The first experiments with the geostationary communication satellite TDRSS in 1986-1988 (Levy et al. 1986) and the first dedicated radioastronomical satellite VSOP (Hirabayashi 2000) were quite successful. The highest lower limit of the brightness temperature, $T > 5.8 \cdot 10^{13}$ K in the rest frame of the IDV source [HB89] 0235+164, was recorded with VSOP at 6 cm wavelength (Frey et al. 2000). VSOP's orbit has a period of 6.3 hours and an apogee radius of 26,000 km.

The RADIOASTRON mission involves the participation of many organizations and many institutes of many countries. A nominal mission orbit has a 28 hour period and the apogee radius is 85,000 km, but an essential increase of the orbit is under strong preparation. The space radiotelescope (SRT) has a reflector antenna of 10 m diameter and will be operated in four wavelength bands, P (92 cm), L (18 cm), C (6.2 cm) and K (1.35 cm), with two circular polarizations for each. Table 2 demonstrates the main parameters of SRT including a fringe size for the different orbits (Kardashev 1997, ASC 2000).

As mentioned before and in connection with new results about the ultracompact structure of many extragalactic radio sources it was decided to start a technological program preparing for the launching on the more distant orbit: the primary orbit has a 9.5 day period, radius apogee 350,000 km and radius perigee 28,000 km. This orbit has a small but regular evolution due to the Moon gravity perturbations, which it is possible to use for the observations of each source in the sky with the largest, medium and even with on-the-ground (for calibration) baseline projections. After a few years of observations with such an orbit it is possible to transfer SRT to an orbit

Table 2. The main RADIOASTRON parameters

Band	P	L	C	K
SPACE RADIOTELESCOPE PARAMETERS				
Frequency (GHz)	0.327	1.66	4.85	22.2
Max recording bandwidth (MHz)	4	32	32	32
System temperature (K)	70	50	50	60
Antenna efficiency	0.3	0.5	0.5	0.3
Tsys (Jy)	8200	3500	3500	7000
GROUND RADIOTELESCOPE PARAMETERS (upgraded VLA, 116 m diameter equivalent reflector)				
System temperature (K)	70	26	21	54
Antenna efficiency	0.5	0.5	0.6	0.5
Tsys (Jy)	37	14	14	28
INTERFEROMETER SENSITIVITY (at 300 sec integration time)				
Coherence factor	1.0	1.0	0.9	0.8
Min. cor. flux (mJy, RMS)	11	1.6	1.7	4.0
FRINGE SIZE AT BASELINE B (microarcseconds)				
B = 85,000 km	2200	440	150	33
B = 350,000 km	540	106	37	8
B = 3.2 million km	59	12	4	0.9

with an apogee radius of 3.2 million km by a gravitation maneuver near the Moon (ASC 1999a).

For the shortest wavelength K band which provides the highest angular resolution it is planned to apply a multi-frequency synthesis method during the image reconstruction (Andreyanov at al. 1984, ASC 1999b). The one spectral channel with fixed K band receiving frequency will be operated simultaneously with the second channel which operates on the 8 fixed and successively switching frequencies in the 18.4 – 25.1 GHz band. Such a system gives essentially improved imaging quality even for short time observations. It is possible to construct a one dimensional image in about 8 times the integration time and to investigate an expansion or proper motion in a time of about one hour. This system permits to operate a satellite space-ground interferometer similarly to the operation of an N-satellite system, where N is the number of fixed channels ($N = 8$ for RADIOASTRON), which orbit synchronously and are distributed along the same orbit radius on the distances from the Earth proportional to the channel frequency. The percentage of the

Fig. 3. RADIOASTRON engineering model

frequency tuning is the portion of the UV coverage for the one dimensional imaging in short time and for two dimensional imaging at the motion along the orbit.

5 Conclusion

There are two systematic targets for space VLBI. One is the improving of the imaging quality in order to obtain the maximum number of details of the image. It is determined by the dynamic range and minimum detectable brightness temperature which is connected mainly with improving the UV coverage and sensitivity of the system and such investigations with a space radio telescope are possible only on the smaller orbits. The second type of targets requires a maximum angular resolution and therefore for a space radio telescope will be connected with the maximum brightness temperature sources. An investigation of the central parts of galactic nuclei constitutes exactly such a type of object. A new orbit for the RADIOASTRON mission is very convenient for the study of the central machines. The most convenient sources for this program are IDV objects, and the main processes of very intensive radio emission are due to cosmic ray synchrotron radiation.

References

1. Andreyanov, V.V., Gurvits, L.I., Kardashev, N.S., Pogrebenko, S.V., Rudakov, V.A., Sagdeev, R.Z., Tsarevsky, G.S. (1984): in Quasat — a VLBI observatory in space, Workshop, Austria, ESA SP-213, p.161
2. ASC (2000): http://www.asc.rssi.ru/radioastron/summary.htm
3. ASC (1999a): http://www.asc.rssi.ru/radioastron/Memo/Dec99/memo_dec99.htm
4. ASC (1999b): http://www.asc.rssi.ru/radioastron/Memo/Sep99/memo_sep99.htm
5. Blandford, R.D., Znajek, R.L. (1977): MNRAS 179, 433
6. Dennett-Thorpe, J., de Bruyn, A.G. (2000): ApJ 529, L65
7. Frey, S., Gurvits, L.I., Altschuler, D.R., Davis, M.M., Perillat, Ph., Salter, Ch.J., Aller, H.D., Aller, M.F., Hirabayashi, H. (2000): astro-ph/0007347
8. Hirabayashi, H. (2000): in Astrophysical Phenomena Revealed by Space VLBI, Proc. of VSOP Symposium, H. Hirabayashi, P.G. Edwards, D.W. Murphy (eds.), ISAS, Japan, p. 3
9. Homan, D.C., Wardle, J.F.C. (1999): AJ 118, 1942
10. Jauncey, D.L., Kedziora-Chudczer, L.L., Lovell, J.E.J., Nicolson, G.D., Perley, R.A., Reynolds, J.E., Tzioumis, A.K., Wieringa, M.H. (2000): in Astrophysical Phenomena Revealed by Space VLBI, Proc. of VSOP Symposium, H. Hirabayashi, P.G. Edwards, D.W. Murphy (eds.), ISAS, Japan, p. 147
11. Jukes, J.D. (1967): Nature 216, 461
12. Kardashev, N.S. (2000): Astronomy Reports (accepted)
13. Kardashev, N.S. (1995): MNRAS 276, 515
14. Kardashev, N.S. (1997): Experimental Astronomy 7, 329
15. Kedziora-Chudczer, L., Jauncey, D.L., Wieringa, M.W., Reynolds, J.E., Tzioumis, A.K., Walker, M.A., Nicolson, G.D. (2000): Adv. Space Res 26, 727
16. Kedziora-Chudczer, L., Jauncey, D.L., Wieringa, M. (1997): ApJ 490, L9
17. Kellermann, K.I., Pauliny-Toth, I.I.K. (1969): ApJ 155, L71
18. Kellermann, K.I., Vermeulen, R.C., Zensus, J.A., Cohen, M.H. (1998): AJ 115, 1295
19. Koopmans, L.V.E., de Bruyn, A.G., Wambsganss, J., Fassnacht, C.D. (2000): astro-ph/0004285
20. Kraus, A., Witzel, A., Krichbaum, T.P. (1999): astro-ph/9902328
21. Levy, G.S., Linfield, R.P., Ulvestad, J.S., Edwards, C.D., Jordan, J.F., Jr., di Nardo, J., Christensen, C.S., Preston, R.A., Skjerve, L.J., Blaney, K.B. (1986): Science 234, 187
22. Macdonald, D., Thorne, K.S. (1982): MNRAS 198, 345
23. Moellenbrock, G.A., Fujisawa, K., Preston, R.A., Gurvits, L.I., Dewey, R.J., Hirabayashi, H., Inoue, M., Kameno, S., Kawaguchi, M., Iwata, T., Jauncey, D.L., Migenes, V., Roberts, D.H., Schilizzi, R.T., Tingay, S.J. (1996): AJ 111, 2174
24. Pacini, F., Rees, M. (1970): Nature 226, 622
25. Preston,R.A., Lister, M.L., Tingay, S.J., Piner, B.G., Murphy, D.W., Meier, D.L., Pearson, T.J., Readhead, A.C.S., Hirabayashi, H., Kobayashi, H., Inoue, M. (2000): in Astrophysical Phenomena Revealed by Space VLBI, Proc. of VSOP Symposium, H.Hirabayashi, P.G. Edwards, D.W. Murphy (eds.), ISAS, Japan, p. 199

26. Qian, S.J., Kraus, A., Witzel, A., Krichbaum, T.P., Zensus, J.A. (2000): A&A 357, 84
27. Quirrenbach, A., Kraus, A., Witzel, A., Zensus, J.A., Peng, B., Risse, M., Krichbaum, T.P., Wegner, R., Naundorf, C.E. (2000): A&AS 141, 221
28. Quirrenbach, A., Witzel, A., Wagner, S., Sanchez-Pons, F., Krichbaum, Krichbaum, T.P. (1991): ApJ 372, L71
29. Rees, M. (1967): MNRAS 135, 345
30. Romero, G.E., Combi, J.A., Benaglia, P., Ascarate, I.N., Cersosimo, J.C., Wilkes, L.M. (1997): A&A 326, 77
31. Romero, G.E., Surpi, G., Vucetich, H. (1995): A&A 301, 641
32. Taylor, G.B. (1998): ApJ 506, 637
33. Wagner, S.J., Witzel, A. (1995): ARA&A 33, 163
34. Wagner, S.J., Witzel, A., Heidt, J., Krichbaum, T.P., Qian, S.J., Quirrenbach, A., Wegner, R., Aller, H., Aller, M., Anton, K., Appenzeller, I., Eckart, A., Kraus, A., Naundorf, C., Kneer, R., Steffen, W., Zensus, A. (1996): AJ 111, 2187
35. Woltjer, L. (1966): ApJ 146, 597

Non-cosmological Redshifts

G. Burbidge

University of California, San Diego

Abstract. The observational evidence showing that many QSOs and related objects have large redshift components which are not attributable to the expanding universe is described and reviewed. It is concluded that an understanding of this phenomenon is fundamental to further investigation of the extragalactic universe.

1 Introduction

I am very happy to help celebrate the 70th birthday of Lo Woltjer. As a friend and contemporary of his I have always enjoyed his company, his insight, and his approach to science and I have been very impressed by his achievements. I first met him at the Yerkes Observatory more than 40 years ago and my first scientific interactions with him were to do with Seyfert galaxies which we (Margaret Burbidge, Kevin Prendergast and I) were working on, and in which he had then become very interested. Over the years he and I have worked on a number of occasions on similar problems, though we have often come to different conclusions! Finally, on a personal level I want to stress how much I believe that Lo has contributed to astronomy through his administrative work and particularly his outstanding period as Director-General at ESO.

Today I want to talk about what I believe is the central problem of extragalactic astronomy, namely whether or not all redshifts can be explained as Doppler shifts and/or cosmological shifts.

2 Redshifts of Galaxies and QSOs

More than 70 years ago Hubble (1929) clearly showed that there was a correlation between redshift and distance for the spiral nebulae, and by using Lemaitre's solutions (Lemaitre 1927) to Einstein's equations (Friedmann's earlier work of 1922 was not known in the west at the time) it was immediately deduced that the universe is expanding. This was probably the most important astronomical discovery of the twentieth century.

From it the very strong belief developed that whenever a redshift is measured it must be cosmological in origin, and relate to distance and lookback time in the universe. As fainter and fainter galaxies were observed, Hubble, Humason, Mayall and Sandage and later others, showed that a smooth redshift apparent magnitude diagram emerged, and it became clear that departure from the linear relation between m and log z which start to appear for

78   G. Burbidge

$z \gtrsim 0.5$ might be able to give us cosmological information (cf Sandage 1961). This existence of the Hubble law for galaxies very strongly supports the view that the redshifts of galaxies of stars are largely cosmological in origin, though the quantization effects for small Δz (cf Tifft 1976; Guthrie and Napier 1996) suggest that we do not understand these redshifts completely.

However, as soon as the first few quasi-stellar objects were discovered in the early 1960s it became clear that they did not follow the simple Hubble law for galaxies which had led to the discovery of the expansion of the universe. The apparent magnitude redshift plot for QSOs is practically a scatter diagram (Fig 1). Of course this does not mean that the redshifts are not cosmological in origin, but it means that if they are there must be a very large spread in absolute magnitude of every redshift. In the same early period of discovery rapid variability of radio and optical flux of the QSOs was discovered. This had never been found in any extragalactic object before. It indicated that the sources are very small, and led to a basic problem in understanding the physics of the radiation process. This was the so-called Compton paradox (Hoyle, Burbidge & Sargent 1966) which could be got around *either* by supposing that the QSOs are much closer than their redshifts suggest, or by appealing to highly relativistic motions (Woltjer 1966; Rees 1967). This latter idea became the preferred solution. Using this approach the discovery of measurable changes of angular size in radio structure on a timescale of years in some compact bright radio emitting QSOs led to the view that highly relativistic motions (superluminal) are present with values of $\gamma = (1 - v^2/c^2)^{-1/2} = 5$ to 10 or even greater. No serious attention was given to the non-cosmological redshift hypothesis as even a partial alternative to the superluminal idea.

But in 1966 Hoyle and I from all of the evidence then available, showed that as good a case could be made for supposing that QSOs were ejected from comparatively nearby galaxies, as could be made by supposing that they lie at the distances indicated by their (cosmological) redshifts. (Hoyle and Burbidge 1966). If they are comparatively local, then the bulk of their redshifts must be intrinsic (non-cosmological) and they cannot be used to probe the distant universe.

The observational test of the non-cosmological redshift hypothesis requires a demonstration that two extragalactic objects, one at least being a QSO, with very different redshifts, are physically associated together in space. We can also ask whether there are properties of QSO redshifts which cannot be explained in terms of an expanding universe. But as far as direct methods are concerned since we have normally *used* the redshifts to measure distance, we can only *test* the cosmological redshift hypothesis by using statistical methods to prove physical association, or try to find luminous connections between objects with different redshifts, or show that the redshift distribution cannot be understood in the framework of cosmological models.

In the last thirty years a large body of observational evidence showing that many QSOs do have large non-cosmological redshift components has built

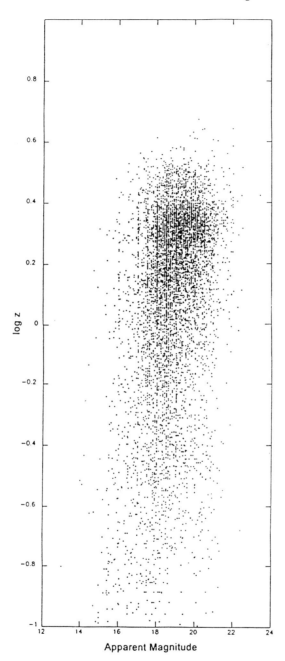

Fig. 1. The Redshift Apparent Magnitude Plot for more than 7000 QSOs. This figure was originally published by Hewitt & Burbidge (1993)

up and by now has become overwhelming. I summarize it in the following sections:

2.1 Statistical Investigations

Here the method is to use well defined samples of QSOs and compare their positions with the positions of bright low-redshift galaxies in published catalogues. The following studies are to be found in the literature:

(i) A comparison of the QSOs in the 3CR catalogue (50 QSOs) with the bright galaxies in the Shapley-Ames Catalogue (about 1200 galaxies). Here 5 of the QSOs were found to lie within 6' of the centers of bright galaxies, and the effect is significant at the 6σ level (Burbidge et al. 1971; Kippenhahn & de Vries 1974)

(ii) A comparison of a QSO sample dominated by radio emitting objects largely identified from the 3CR, Molonglo, Parkes and 4C radio catalogues (about 1500 QSOs) with the bright galaxies in the de Vaucouleurs catalogue (about 3400 galaxies with $z \leq 0.05$) gave a highly significant result (Chu et al. 1984)

(iii) Similarly, strong correlations of high redshift QSOs have been found with galaxies in the IRAS catalogue where a few galaxies have values of z up to 0.4, but most have very much lower redshifts. The QSOs are a complete sample with radio flux ≥ 1 Jansky (Stickel, Fried & Kuhr 1993; Bartelmann & Schneider 1993, 1994)

(iv) A very strong correlation has also been found between the Large Bright QSO Survey QSOs in the field of the Virgo cluster (178 QSOs) and the Virgo cluster galaxies in the Binggeli, Tammann, Sandage catalogue (Binggeli et al. 1987) by Zhu and Chu (1995).

(v) Strong correlations on scales $\leq 10'$ have been found between optically bright high redshift radio loud QSOs and the diffuse x-ray emitting sources detected by ROSAT. Bartelmann et al. (1994) believe that the diffuse x-ray sources are due to galaxy clusters with redshifts significantly less than the QSO redshifts.

(vi) Strong correlations have been found between the high redshift radio QSOs in the sample of Stickel, Fried & Kuhr, and the galaxies in the Lick catalogue ($z \leq 0.2$) (Shane & Wirtanen 1967) (Bartelmann & Schneider 1993, 1994; see also Seldner & Peebles 1979). The significance level is up to 98%.

(vii) Stocke et al. (1987) using an early sample of x-ray emitting QSOs showed that their associations with moderate redshift galaxies ($z \leq 0.15$) was statistically significant at a high confidence level ($> 97.5\%$).

Is there any possible way in which the close proximity of QSOs with high redshifts and galaxies with low redshifts could be understood while still assuming that they both have cosmological redshifts? The only possible mechanism is weak gravitational lensing which was invoked by Bartelmann et al. (1994) in their studies of faint clusters and high redshift QSOs. They supposed that large amounts of dark matter are present in all of the clusters of galaxies which show strong positional correlations with much higher redshift QSOs. This idea has been taken up by several other groups who have found similar correlations more recently (Seitz & Schneider 1995; Norman & Williams 2000; Williams & Irvine 1998).

All of these groups show that the correlations are real. They all show their bias since none of them even mentions the possibility that this is further evidence for non-cosmological redshifts, even though they find that the amplitude of the correlations is a significant factor less than the lensing models predict (Williams 2000). In fact on a quantitative basis it is clear that the models won't work. And of course the correlations found involving the very closeby galaxies, in the Shapley-Ames catalogue or the Virgo cluster galaxies, cannot be explained in this way. Thus, all of these studies provide strong evidence that the galaxies are *physically associated* with the QSOs so that the QSO redshifts must have large components which are not of cosmological origin.

2.2 Associations of Individual QSOs with Bright Galaxies

We now turn to the many individual cases in which QSOs are found to lie very close to bright low-redshift galaxies. The general point to make here is that QSOs are very rare objects when compared with galaxies, and we do not expect to see very many accidental configurations even if all ~ 10000 bright galaxies known were surveyed. And they have not been. In 1996 when I last attempted to compile all of the close pairs consisting of a bright nearby galaxy and a high redshift QSO which are known, there were 47 pairs, 44 of them with separations $\leq 3'$ (Burbidge 1996, Table 1). All of the galaxies have $m \leq 15.5$. Two of these galaxies, NGC 470 and NGC 622 have 2 associated QSOs and two more, NGC 1073 and NGC 3842 have 3. The field around NGC 3842 is shown in Figure 2. Since we have catalogues of all of the bright galaxies in the sky, and also from surveys of limited areas, we have good estimates of the surface density of QSOs as a function of magnitude (cf Goldschmidt et al. 1992, Boyle et al. 1990), the probability that any close pair is a result of an accidental configuration can be calculated in a straightforward way. We suppose that the QSOs are distributed randomly on the sky. Then if the number of pairs we expect to find by chance with separation of θ arc minutes is n, then

$$n = 8.64 \times 10^{-4} \Gamma \theta^2 N \qquad (1)$$

where Γ is the surface density of QSOs and N is the number of cases that have been investigated.

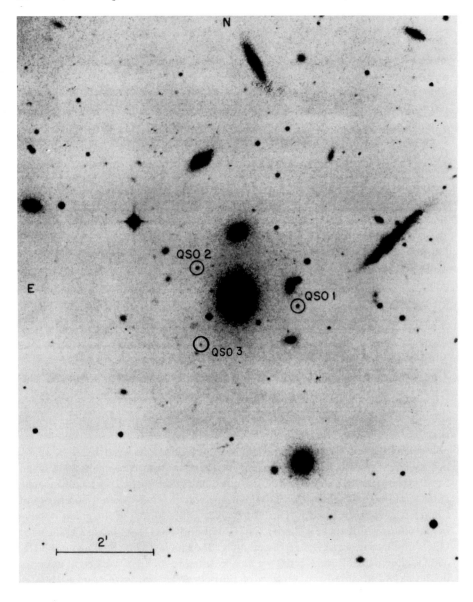

Fig. 2. Three QSOs with redshifts 0.335, 0.946 and 2.205 very close to NGC 3842. Two of them are x-ray sources (Arp 1987)

Of the 47 pairs mentioned above, a few were found serendipitously, and some through the early work of Burbidge et al. (1971). Apart from these studies only Arp looked carefully at a fraction of the fields around bright galaxies and he only examined ≲ 200 before his observing program was abruptly stopped. A comparison between the number of pairs expected by accident, and those actually found, is shown in Table 1. Here we have very conservatively assumed that the number of galaxy fields searched is $N = 500$. There has still never been a systematic attempt made to identify QSOs near to the ~ 10000 galaxies brighter than about 15.5. Table 1 provides very strong evidence that the galaxies and adjacent QSOs are not accidental pairings. In addition, the luminous connections found in a few cases, and notable in the pair NGC 4319 - Markarian 205 (Sulentic & Arp 1987) also demonstrate the physical association between active galaxies and QSOs with very different redshifts.

Table 1.

m(QSO)	QSO Surface Density (No./square degree)	No. of pairs expected by chance	No. of pairs actually found
≤ 16.5	0.022	0.1	3
≤ 17.5	0.1	0.4	10
≤ 18.5	1	4	26

Is there any way to explain the large number of close QSO-galaxy pairs and still preserve the cosmological redshift hypothesis? Canizares (1981) suggested that if the halos of the galaxies were made up of large numbers of dark stars, gravitational micro-lensing of faint background QSOs might explain the effect. But it has been shown by several authors, Schneider (1994), Arp (1990) and Ostriker (1989), that the surface density of faint QSOs which is required to explain the observed number of pairs is far in excess of the observed density. Moreover, gravitational lensing cannot explain luminous connections, or alignments of QSOs or redshift periodicities, all of which are sometimes present.

2.3 The Connection Between Active Galaxies and X-Ray Emitting QSOs

We already pointed out that Stocke et al. (1987) showed that there was clearly a statistical association between moderate redshift galaxies and x-ray emitting QSOs. In 1997 Radecke (1997) using the ROSAT Survey of compact x-ray sources showed that the compact x-ray sources tend to cluster about active galaxies. The significance of the association ranges as high as 7.4σ for sources lying at distances between 10′ and 40′ from the galaxies. It was further shown (Arp 1997, Pietsch et al. 1994) that many of these sources could be identified with blue stellar objects which turn out to be high redshift QSOs (Burbidge 1995, 1997, 1999, 2000; Chu et al. 1998). In Table 2 we give a list of the galaxies in which clustered x-ray emitting QSOs have been identified and the redshifts of the QSOs, and in Figs. 3–5 we give some examples of the configurations. The statistical evidence combined with the morphology of the kind seen in the figures very strongly suggests that the QSOs are ejected from the galaxies, often along a preferred axis.

Table 2. Redshifts of all x-ray emitting QSOs so far observed aligned across Seyfert/AGN galaxies

Galaxy	z_{gal}	QSO Redshifts (z_{obs})					
NGC 1068	0.004		0.261	0.655			
NGC 2639	0.011		0.305 0.323				
NGC 3079	0.0038		0.216		1.022	1.405	
NGC 3516	0.0087		0.328	0.690	0.929	1.399	2.101
NGC 3628	0.003				0.983		
NGC 4258	0.0015		0.398	0.653			
NGC 4579	0.005	0.106		0.662	0.947		
Mkn 231	0.041		0.320	0.489			
Mkn 273	0.038		0.376	0.600	0.941 1.163		
NGC 5273	0.0035		0.33				
NGC 5548	0.017		0.184	0.727 0.560 0.674	0.852		
NGC 5689	0.0076					1.358	1.94 2.391
IC 4553	0.018			0.459			
NGC 6217	0.005		0.358 0.376 0.380		1.134		
Peaks found in earlier total distribution		0.061	0.30	0.60	0.96	1.41	1.96

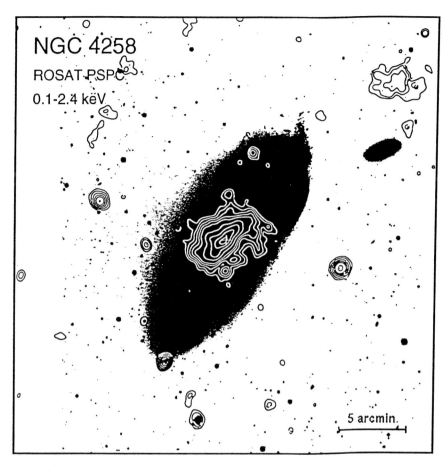

Fig. 3. Two x-ray emitting QSOs apparently ejected from NGC 4258 (Table 2) which is only 7.2 Mpc away (Pietsch et al. 1994; Burbidge 1995)

3 Peaks and Periodicities in the Redshift Distribution of QSOs

There is one final class of evidence to be described. This evidence comes from the existence of peaks and periodicities in the distribution of observed redshifts of QSOs. So far I have shown from statistical and morphological evidence that some parts of the redshifts of the QSOs are not due to the cosmological expansion. Let us call this part of the redshift the intrinsic component z_i. Then the observed redshift z0 is given by

$$(1 + z_0) = (1 + z_c)(1 + z_d)(1 + z_i) \quad (2)$$

where z_c and z_d are the cosmological and Doppler components respectively.

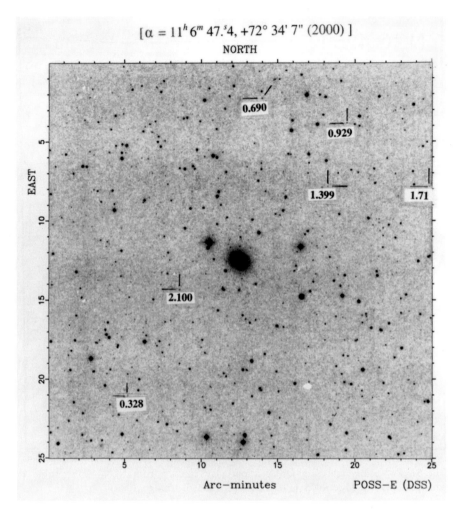

Fig. 4. Six QSOs, five of them being x-ray sources associated with NGC 3516 (Table 2) (Chu et al. 1998)

We see that if $z_i = 0$, peaks and periodicity in the observed redshift distribution would only be seen if z_c were responsible. But there is no evidence that such an effect is present in the spectra of normal galaxies. Thus if peaks in the QSO redshift distribution are seen, we are forced to the conclusion that $z_i \neq 0$ and it is responsible for the effect. Then to see peaks at all, $z_0 \approx z_i$. For this to occur, $z_c \ll 1$, and $z_d \ll 1$. For galaxies in general the random motions indicate that $z_d \leq 0.001$. Thus z_d is expected to be very much less than 1. However, a range of values of z_c from 0.001 to 0.05, say, which corresponds to a modest cosmological distance range, would completely smear out any observed peaks due to z_i even if they were present. Thus the *appearance* of

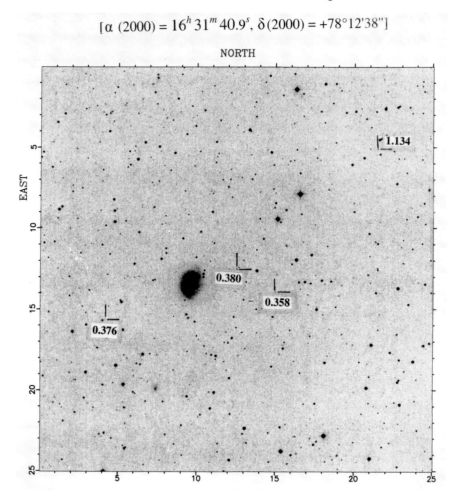

Fig. 5. Four x-ray emitting QSOs associated with NGC 6217 (Table 2) (Burbidge 2000)

sharp peaks would force us to the conclusion that the QSOs with redshifts in or near the peaks must be comparatively local objects. For QSOs in the general field which may well have a range of cosmological redshifts, any peaks will be smeared out. Thus, to see a peak at all would be remarkable. However, in 1967 a peak in the observed redshift distribution for QSOs was found at $z_0 = 1.955$, and in 1968 a second peak involving both QSOs and related objects (now called AGN) was found at 0.061 (Burbidge & Burbidge 1967; Burbidge 1968). By the 1970s several more peaks had appeared, at $z = 0.30$, 0.60, 0.96 and 1.41, and Karlsson (1971) showed that the peaks could be fitted to a periodic sequence such that $\Delta \log(1 + z) = 0.089$.

I show in Fig. 6 a histogram of the redshift distribution dating from 1977. The QSOs are very largely radio-emitting QSOs which clearly show the peaks at 0.30, 0.60, 0.96, 1.41 and 1.96. The peak at 0.061 is dominated by objects that are often classified as AGN and not QSOs and thus it was not included in that plot. The existence of a huge peak at this value was confirmed by Burbidge and Hewitt (1990) who analyzed more than 500 AGN and QSOs with small z (Hewitt & Burbidge 1991).

Following their discovery, the reality of the peaks was questioned on two grounds:

1. It was argued that they were a result of spectroscopic selection (cf Wills & Ricklefs 1976) and,
2. It was claimed that other samples did not show the effect (Green & Richstone 1976; Scott 1991)

Both of these objections were easily overcome. It was shown that the shifting of the key emission lines as the redshifts changed could not be doing this (Burbidge 1978, Depaquit et al. 1985). It is now clear that the peaks found in the earlier data (circa 1977) appeared *because* the QSOs came from radio samples where there were no optical selection effects working. If optical selection effects are present they will have an effect on the redshift distribution seen, as is the case, for example in the large bright QSO survey where the periodicity does not appear.

As far as the "new" samples of Green and Richstone and of Scott were concerned, they were actually taken from lists of emission-line galaxies that simply did not fulfill the criteria originally specified – namely that the spectra were made up of the broad emission lines and non-thermal continua which originally defined the energy distribution of the QSOs. The point is that the light is not coming from stars, or gas excited by stars, in the objects in which the redshifts show peaks.

Karlsson (1990) showed that the periodicity previously discovered is found in QSOs which from statistical evidence are associated with bright galaxies. Recently we have found two new samples of QSOs in which we can test the hypothesis (Burbidge & Napier 2000). The Karlsson formula enables us to predict that redshift peaks beyond 1.96 should appear, at $z = 2.63, 3.44, 4.47, 5.71$ etc. if enough high redshift QSOs are contained in the sample. These samples are chosen to avoid any optical selection effects. The first is the sample of x-ray emitting QSOs which we previously discussed (Table 2).

The histogram of that distribution is shown in Fig 7, and it is obvious that this new sample shows the same effect. Since for each system in Table 2 we know the redshift of the galaxy, if we suppose that in each QSO the intrinsic redshift component is exactly the value in the nearest peak, from equation (2) we can calculate z_d. For the sample we find that $|cz_d| \simeq 12500$ km sec^{-1} suggesting that the QSOs are being ejected from the galaxies at speeds $\sim 0.03 - 0.1c$.

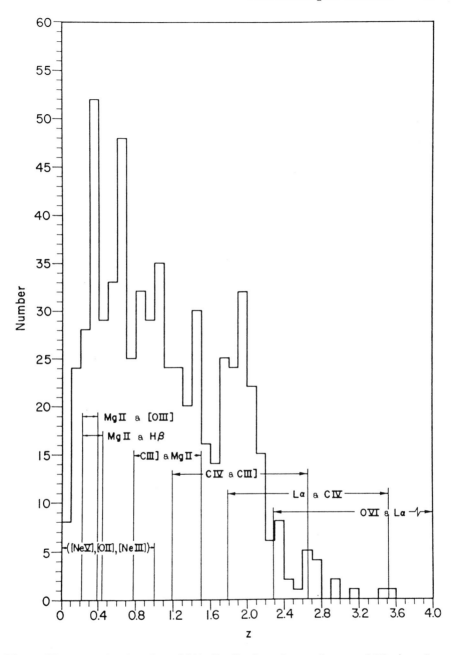

Fig. 6. Histogram showing the redshift distribution of more than 600 QSOs (mostly identified from their radio properties) known in 1977. The redshift peaks at $z = 0.30, 0.60, 0.96$ and 1.41 are clearly visible. The redshift ranges over which different emissions are important are marked. This plot showed that the peaks were not due to spectroscopic selection effects

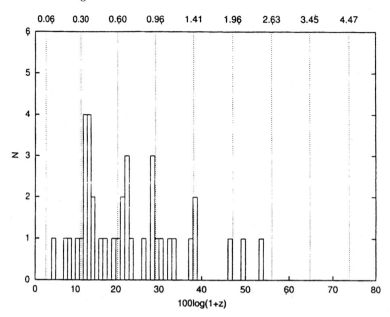

Fig. 7. Histogram of the redshift distribution of x-ray emitting QSOs associated with active galaxies (Table 2), plotted in units of $100\log(1+z)$. The vertical dotted lines represent the positions of the previously established peaks (Burbidge & Napier 2000)

The second sample contains all of the QSOs which are found to be very close double or multiple systems. They are chosen on purely morphological grounds. As was the case for the first sample, nearly all of these QSOs have been found since the original periodicity was identified more than twenty years ago. Here the QSOs are chosen because of their morphological characteristics which have nothing to do with their spectroscopic properties, and they cover a much larger range in redshift than was available earlier. There are 57 redshifts involving objects which have separations $\leq 10''$ in this sample. The close pairs and multiples are supposed conventionally to be gravitational lenses (if the separate images have identical properties), binary quasars (if the separate images have the same redshifts but slightly different optical or radio properties) or accidents, (if the redshifts are very different). All of the known cases are included. They are taken from the compilation of Kochanek et al. (The Castles Survey), from Table 5 of Véron and Véron (2000) and from Burbidge, Hoyle and Schneider (1997). The histogram of the redshift distribution is shown in Fig. 8. We see that the peaks at $z = 2.63$, 3.44 and 4.47 which were predicted, actually appear. The combined histogram from the two samples is shown in Fig. 9. Burbidge and Napier (2000) have done detailed statistical studies of these two samples and have shown that the probability that this is an accidental effect is extremely small, about 10^{-5}.

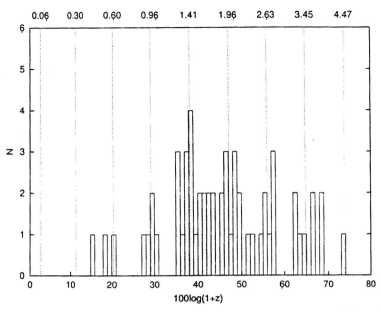

Fig. 8. Histogram of the redshift distribution of close pairs and multiple QSOs. The same units are used as those in Fig. 7

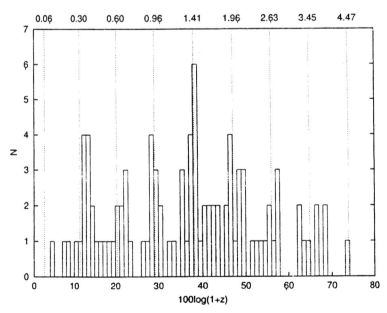

Fig. 9. A combined histogram for the two independent samples given in Figures 7 and 8 (Burbidge & Napier 2000)

4 The Arguments Against Non-cosmological Redshifts

Evidence or interpretations of evidence which are sometimes brought up by those who do not want to believe these conclusions are as follows:

(1) Some QSOs are embedded in galaxies made up of normal stars, and the redshift of the fuzz (the putative normal galaxy) is the same as that of the QSO. The best evidence of this kind comes from the work of Miller (2000) who has found a few cases of this kind. Of course this does not tell us that the non-cosmological redshift evidence is incorrect, but it does tell us that z_i can range from values close to zero to the very large values described previously. The imaging studies suggest that many QSOs may lie in galaxies, but there is a variety of morphologies and for these no spectroscopic evidence that starlight with the same redshift as the QSO light is present, has been found.

(2) The absorption spectra of QSOs are usually *assumed* to be due to intervening matter. While large observing programs are based on that assumption, there is no objective proof that this assumption is correct. For many QSOs which are comparatively closeby, the absorption must be intrinsic as it is accepted to be the case for the broad absorption line QSOs (BAL), and recently it has been concluded that many of the sharp absorption-line systems may be intrinsic to the QSOs (Richards et al. 1999). A detailed discussion of attempts to reconcile the absorption line interpretation with the existence of non-cosmological redshifts was given by Burbidge (1996). In some cases it may very well be that high redshift QSOs are ejected from parent galaxies (damped $Ly\alpha$ systems) which are themselves at large cosmological redshifts, so that some of the absorption may be due to intervening gas.

(3) The gravitational lens interpretation of multiple image QSOs requires them to be at cosmological distances. However, the fact that the periodicity effect is seen in a sample of QSOs *containing* many claimed lensed systems argues against that popular interpretation. Moreover in the sample studied here there are many which are claimed to be binary QSOs for which of course there is no observational argument for putting them at great distances, and close pairs with very different redshifts which on statistical grounds give further evidence for non-cosmological redshifts (cf the arguments given by Burbidge et al. 1997). Also it should not be forgotten that the prototype gravitational lens pair 0957+561A,B has an observed redshift $z = 1.41$, *exactly* on one of the redshift peaks, is an x-ray emitting QSO, and lies only 15' from the active galaxy NGC 3079 (Table 1). This is never mentioned by the believers in gravitational lenses.

(4) Many years ago it was claimed (Setti & Woltjer 1979, 1989) that if the QSOs were local, they would give rise to an x-ray background flux

much larger than that observed. However, Burbidge & Hoyle (1998) have shown that if QSOs are ejected from spiral galaxies they must reflect the space density of normal spiral galaxies as a function of z_c and the majority of the resolved discrete sources which contribute to the background will have $z_c \simeq 0.1 - 0.2$ whatever their observed redshifts are. A specific calculation based on the ROSAT data then showed that there is no x-ray background problem.

5 Conclusion

All of the observational evidence described in sections 2, 3 and 4 clearly points to one major conclusion. There exist objects which have large redshift components which are not associated with the expanding universe. They are all either QSOs or objects closely related to them. The intrinsic redshift component z_i can range all the way from zero in some cases to the very large values described earlier.

Whatever is said or done the evidence described here cannot be avoided. It tells us that non-cosmological redshifts do exist, maybe not in all QSOs, and that they show periodic effects. They need to be understood - Hoyle and I have attempted a first explanation (Hoyle & Burbidge 1996). Almost certainly the major astrophysical problem as we begin this century is to understand this phenomenon. Since much of the observational evidence is in the literature we may ask why it is that it is being ignored. Here the problem appears to be sociological – not scientific.

Early on in the studies of QSOs and radio sources the community showed by its response to the work of Arp and others that it was not prepared to treat the evidence on its merits (cf Arp 1987, etc.). This is even more true today. Apparently for most astronomers so much rests on the idea that redshifts must always measure distance. Very large redshifts and very large energies are apparently very attractive. Current observational programs concerning the early universe are based on the *assumption* that we can use QSOs as cosmological probes. This evidence summarized here shows that we cannot. Also the existence of large non-cosmological redshifts strongly suggests that we may have to learn something new about physics from astronomy, and in modern times this idea is anathema to theorists.

It is not surprising therefore that the few astronomers who believe that there is a case for non cosmological redshifts cannot get research support or telescope time, or that young people are afraid to touch the subject. How long this state of affairs will last I do not know. But there are precedents. The earlier one in the last century that I usually quote is the sad story of Wegener and continental drift.

But one can go back much further. I end with a letter from Galileo Galilei to one of his patrons which might have been written today by any one of the small group of astronomers who dares to hold to the minority position.

"As your Serene Highness well knows, I discovered in the heavens many things that have not been seen before our own age. The novelty of these things, as well as some consequences which followed from them in contradiction to the physical notions commonly held among academic philosophers, stirred up against me no small number of professors – as if I had placed these things in the sky with my own hands in order to upset Nature and overturn the sciences. They seemed to forget that the increase of known truths stimulates the investigation, establishment and growth of the arts; not their diminution or destruction.

Showing a greater fondness for their own opinion than for truth, they hurled various charges and published numerous writings filled with vain arguments, and they made the grave mistake of sprinkling them with passages taken from places in the Bible which they had failed to understand properly and which were ill suited to their purposes."

(Extract of a letter written in 1609 by Galileo Galilei to the Grand Duchess Cristina of Lorraine, devout Catholic, wife of Grand Duke Ferdinando I of Florence)

References

1. Arp, H.C. 1990, A&A, 229, 93
2. Arp, H.C. 1997, A&A, 316, 57
3. Arp, H.C. 1987, Quasars, Redshifts & Controversies, (Berkeley Interstellar Media)
4. Bartelmann, M. & Schneider, P. 1993, A&A, 271, 421
5. Bartelmann, M. & Schneider, P. 1994, A&A, 284, 1
6. Bartelmann, M., Schneider, P. & Hasinger, G. 1994, A&A, 290, 399
7. Binggeli, B., Tammann, G., & Sandage, A. 1987, AJ, 94, 251
8. Boyle, B., Shanks, T. & Peterson, B.A. 1990, MNRAS, 243, 1
9. Burbidge, E.M. 1995 A&A, 298, L1
10. Burbidge, E.M. 1997, ApJ, 477, L13
11. Burbidge, E.M. 1999, ApJ, 511, L11
12. Burbidge, E.M. 2000, Private Communication
13. Burbidge, E.M. & Burbidge G. 1967, ApJ, 148, L107
14. Burbidge, G. 1968, ApJ, 154, L1
15. Burbidge, G. 1978, Physica Scripta, 17, 237
16. Burbidge, G. 1996, A&A, 309, 9
17. Burbidge, G. & Hewitt, A. 1990, ApJ, 359, L33
18. Burbidge, G. Hoyle, F. & Schneider P. 1997, A&A, 320, 8
19. Burbidge, G. & Hoyle, F. 1998, MPE Report, 263, p.307
20. Burbidge, G. & Napier, W. 2000, AJ, in press
21. Burbidge, M., Burbidge G., Solomon, P. & Strittmatter, P. 1971, ApJ, 170, 223
22. Chu, Y., Zhu, X., Burbidge, G. & Hewitt, A. 1984, A&A, 138, 408
23. Chu, Y., Wei, J., Hu, J, Zhu, X., & Arp, H.C., 1998, ApJ 500, 596
24. Canizares, C. 1981, Nature, 291, 620

25. Depaquit, S., Pecker, J.C. & Vigier, J-P., 1985, Astron. Nach., 306, 7
26. Goldschmidt, P., Miller, L., La Franca, F., and Cristiani, S., 1992, MNRAS, 256, 65P
27. Green, R. & Richstone, D. 1976, ApJ, 208, 639
28. Guthrie, B.N., & Napier, W. 1996, A&A, 310, 353
29. Hewitt, A. & Burbidge, G. 1991, ApJ.S., 75, 297
30. Hewitt, A. & Burbidge, G. 1993, ApJ.S., 87, 451
31. Hoyle, F., Burbidge, G. & Sargent, W.L.W. 1966, Nature, 209, 751
32. Hoyle, F. & Burbidge, G. 1966, ApJ, 144, 534
33. Hoyle, F. & Burbidge, G. 1996, A&A 309, 335
34. Hubble, E. 1929, Proc. Nat. Acad. Sci., 15, 168
35. Karlsson, K.G. 1971, A&A, 13, 333
36. Karlsson, K.G. 1990, A&A, 239, 50
37. Kochanek, C.K., Falco, E., Impey, C., Lehar, B., McCleod, R., & Rix, S. 2000, http://cfa-www.harvard.edu/castles
38. Kippenhahn, R. & de Vries, H. 1974, Astrophys. Space Science, 26, 131
39. Lemaitre, G. 1927, Ann. de la Societe Scientifique de Bruxelles, 47, 49
40. Miller, J. 2000, Private Communication
41. Nicholson, K., Mittaz, J. & Mason, K. 1997, MNRAS, 285, 831
42. Norman, D. & Williams, L. 2000, AJ, 119, 2060
43. Ostriker, J. 1989, in BL Lac Objects: Lecture Notes (ed L. Maraschi; Berlin, Springer Verlag) Vol. 334
44. Pietsch, W., Vogler, A., Kahabka, P., Jain, A. & Klein, V. 1994, A&A, 284, 386
45. Radecke, H.D. 1997, A&A, 319, 18
46. Rees, M. 1967, MNRAS, 135, 345
47. Richards, G., Kollgaard, R., Laurent-Muehleisen, S., Vanden Berk D., York, D., & Yanny, B., 1999, ApJ, 513, 576
48. Sandage, A. 1961, ApJ, 133, 355
49. Schneider, P. 1994, Gravitational Lenses in the Universe. eds. J. Surdej et al. (Liege: Institut d'Astrophysique) p. 41
50. Scott, D. 1991, A&A, 242, 1
51. Seitz, S. & Schneider, P. 1995, A&A, 302, 9
52. Seldner, M. & Peebles, P.J. 1979, ApJ 227, 30
53. Setti, G-C. & Woltjer, L. 1979, A&A , 76, L1
54. Setti, G-C. & Woltjer, L. 1989, A&A, 224, L21
55. Shane, C.D. & Wirtanen, C.A. 1967, Pub. Lick. Obs. 22 Part I
56. Stocke, J., Schneider, P., Morris, S. Gioia, I., Maccacaro, T., and Schild, R. 1987, ApJ, 315, L11
57. Stickel, M., Fried, J., & Kuhr H. 1993, A&A, 97, 483; 100, 395
58. Sulentic, J. & Arp, H.C. 1987, ApJ, 319, 687
59. Tifft, W. 1976, ApJ, 206, 38
60. Véron M. & Véron, P. 2000, ESO Scientific Report No. 19
61. Williams, L. & Irvine, M. 1998, MNRAS, 298, 378
62. Williams, L. 2000, ApJ, 535, 37
63. Wills, D. & Ricklefs, R. 1976, MNRAS, 175, 81P
64. Woltjer, L. 1966, ApJ, 146, 597
65. Zhu, X-F. & Chu, Y-Q. 1995, A&A, 297, 300

Supermassive Black Holes

F.D. Macchetto*

Space Telescope Science Institute, 3700 San Martin Drive, Baltimore, MD 21218, USA

Abstract. One of the important topics of current astrophysical research is the role that supermassive black holes play in shaping the morphology of their host galaxies. There is increasing evidence for the presence of massive black holes at the centers of all galaxies and many efforts are directed at understanding the processes that lead to their formation, the duty cycle for the active phase and the question of the fueling mechanism. Related issues are the epoch of formation of the supermassive black holes, their time evolution and growth and the role they play in the early ionization of the Universe. Considerable observational and theoretical work has been carried out in this field over the last few years and I will review some of the recent key areas of progress.

1 Introduction

It is now widely accepted that quasars (QSOs) and Active Galactic Nuclei (AGN) are powered by accretion onto massive black holes. This has led to extensive theoretical and observational studies to elucidate the properties of the black holes, the characteristics of the accretion mechanisms and the mechanisms responsible for the production and transportation of the energy from the central regions to the extended radio lobes.

However, over the last few years there has been an increasing realization that Massive Dark Objects (MDOs) may actually reside at the centers of *all* galaxies [1–7]. The mean mass of these objects, of order $10^{-2.5}$ times the mass of their host galaxies, is consistent with the mass in black holes needed to produce the observed energy density in quasar light if we make reasonable assumptions about the efficiency of quasar energy production [8,9]. This raises a number of important new questions and has fundamental implications for the role of the black holes in contributing or being responsible for the ionization (or reionization) of the early universe and for their role in the processes leading to the formation of galaxies. Conversely the apparent correlation between the black hole mass and the mass of the spheroidal component in elliptical and spirals points towards a close interaction between the galaxy size and morphology and its central black hole. Models in which elliptical galaxies form from the mergers of disk galaxies whose bulges contain black holes are consistent with the "core fundamental plane, the relation between the central

---

* On assignment from the Astrophysics Division, Space Science Department of the European Space Agency

parameters of early-type galaxies. Furthermore it is clear that the dynamical influence of a supermassive black hole can extend far beyond the nucleus if a substantial number of stars are on orbits that carry them into the center. Work by Merritt [10] has shown that nuclear black holes are important for understanding many of the large-scale properties of galaxies, including the fact that elliptical galaxies come in two, morphologically-distinct families, the absence of bars in most disk galaxies, and the shapes of the spiral galaxy rotation curves. Since the growth of the black hole mass depends on the global morphology of the host galaxies, the link between black hole and galaxy structure implies a feedback mechanism that determines what fraction of a galaxys mass ends up in the center.

2 Dynamical Evidence for Massive Black Holes

2.1 Megamasers in NGC 4258

The best observation showing the presence of a Keplerian disk around a black hole was the VLBI observation of megamasers in the nucleus of the Seyfert 2 galaxy NGC 4258 reported by Miyoshi et al. [11]. These observations reveal individual masing knots revolving at distances ranging from ~ 13 pc to 25 pc around the central object. These data show a near-perfect Keplerian velocity distribution, implying that almost all the mass is located well within the inner radius where the megamasers reside, and they derive a central mass of $\sim 3.6 \times 10^7\,M_\odot$ within the inner ~ 13 pc.

Given this mass and the fact that NGC 4258 is a relatively low luminosity ($\sim 10^{42}\,\mathrm{erg\,s^{-1}}$) object, the emission is sub-Eddington, with $L/L_\mathrm{E} \sim 3 \times 10^{-4}$. Such sub-Eddington sources are likely to have accretion disk structures, where the accreting gas is optically thin and radiates inefficiently, and the accretion energy that is dissipated viscously, is advected with the accretion flow (see, e.g. [12–14]).

2.2 Kinematic Studies Using Optical Emission Lines

The other line of evidence for the presence of black holes in galaxies is the velocity field of the matter emitting closely to the nucleus. Very high spatial resolution observations using the long-slit spectrograph on the *Faint Object Camera* of *HST* were carried out by Macchetto et al. [15]. We observed the ionized gas disk in the emission line of [OII]$\lambda 3727$ at three different positions separated by 0.2 arcsec, with a spatial sampling of 0.03 arcsec (or ~ 2 pc at the distance of M87), and measured the rotation curve of the inner $\sim 1''$ of the ionized gas disk to a distance as close as $0\rlap{.}''07$ ($\simeq 5$ pc) to the dynamical center. We modeled the kinematics of the gas under the assumption of the existence of both a central black hole and an extended central mass distribution, taking into account the effects of the instrumental PSF, the intrinsic luminosity distribution of the line, and the finite size of the slit. We found that the central mass must be concentrated within a sphere whose maximum radius is $\simeq 3.5\,\mathrm{pc}$ and showed that both the observed rotation curve and line profiles are best explained by a thin-disk in Keplerian motion. Finally, we proved that the observed motions are due to the presence of a super-massive black-hole and derived a value of $\mathrm{M_{BH}} = (3.2 \pm 0.9) \times 10^9\,\mathrm{M_\odot}$ for its mass.

2.3 Virial Masses

The virial masses and emission-line region sizes of Active Galactic Nuclei (AGNs) can be measured by "reverberation-mapping" techniques. Wandel, Peterson & Malkan [16] have compiled a sample of 17 Seyfert 1 and 2 quasars with reliable reverberation and spectroscopic data and used these results to calibrate similar determinations made by photoionization models of the AGN line-emitting regions. Reverberation mapping uses the light travel-time delayed emission-line response to continuum variations to determine the size and kinematics of the emission-line region. The distance of the broad emission-line region (BLR) from the ionizing source is then combined with the velocity dispersion, derived from either the broad Hβ line width or from the variable part of the line profile to estimate the virial mass. When they compare the central masses calculated with the reverberation method to those calculated using a photoionization (Hβ line) model, they find a nearly linear correlation (Table 1). They find that the correlation between the masses is significantly better than the correlation between the corresponding BLR sizes calculated by the two methods, which further supports the conclusion that both methods measure the mass of the central black hole. They also derive the Eddington ratio, which for the objects in the sample fall in the range $L_V/L_{\mathrm{Edd}} \sim 0.001\text{--}0.03$ and $L_{\mathrm{ion}}/L_{\mathrm{Edd}} \approx 0.01\text{--}0.3$.

2.4 The Black Hole Mass of a Seyfert Galaxy

In a recent study Winge et al. [17] have analyzed both ground-based, and *HST/FOC* long-slit spectroscopy at subarcsecond spatial resolution of the narrow-line region (NLR) of NGC 4151. They found that the extended emission gas ($R > 4''$) is in a normal rotation in the galactic plane, a behavior that they were able to trace even across the nuclear region, where the gas is strongly disturbed by the interaction with the radio jet and connects smoothly with the large-scale rotation defined by the neutral gas emission. The *HST* data, at $0''\!.03$ spatial resolution, allow for the first time truly to isolate the kinematic behavior of the individual clouds in the inner narrow-line region. They find that, underlying the perturbations introduced by the radio ejecta, the general velocity field can still be well represented by planar rotation down to a radius of $\sim 0''\!.5$ (30 pc), the distance at which the rotation curve has its turnover. The most striking result that emerges from the analysis is that the galaxy potential derived fitting the rotation curve

Table 1. Reverberation BLR Sizes and Central Masses Compared with Photoionization Sizes and Masses. The last two columns give the ionizing luminosity derived from the lag and the corresponding Eddington ratio [16]

Name	$\log R_{\rm ph}$	log lag	$\log M_{\rm ph}$	$\log M_{\rm rev}$	$M_{\rm rev}(10^7\,M_\odot)$	$\log L_{\rm ion}$	$\log\left(\frac{L_{\rm ion}}{L_{\rm Edd}}\right)$
3C 120	0.92	1.64	6.86	7.49	$3.1^{+2.0}_{-1.5}$	45.03	-0.57
3C 390.3	0.89	1.38	8.26	8.59	39.1^{+12}_{-15}	44.51	-2.19
Akn 120	1.07	1.59	7.97	8.29	$19.3^{+4.1}_{-4.6}$	44.92	-1.48
F9	1.13	1.23	8.03	7.94	$8.7^{+2.6}_{-4.5}$	44.21	-1.84
1C 4329A	<0.56	0.15	7.34	<6.86	<0.73	<42.04	<-2.93
Mrk 79	0.81	1.26	7.48	8.02	$10.5^{+4.0}_{-5.7}$	44.26	-1.87
Mrk 110	0.78	1.29	6.46	6.91	$0.80^{+0.29}_{-0.30}$	44.33	-0.69
Mrk 335	0.89	1.23	6.68	6.58	$0.39^{+0.14}_{-0.11}$	44.20	-0.49
Mrk 509	1.08	1.90	7.17	7.98	$9.5^{+1.1}_{-1.1}$	45.54	-0.54
Mrk 590	0.85	1.31	7.00	7.15	$1.4^{+0.3}_{-0.3}$	44.37	-0.89
Mrk 817	0.86	1.19	7.53	7.56	$3.7^{+1.1}_{-0.9}$	44.13	-1.54
NGC 3227	0.16	1.04	6.92	7.69	$4.9^{+2.7}_{-5.0}$	43.82	-1.98
NGC 3783	0.52	0.65	7.05	7.04	$1.1^{+1.1}_{-1.0}$	43.05	-2.10
NGC 4051	<0.14	0.81	5.37	6.15	$0.14^{+0.15}_{-0.09}$	41.57	-0.84
NGC 4151	0.44	0.48	7.35	7.08	$1.2^{+0.8}_{-0.7}$	42.70	-2.49
NGC 5548	0.73	1.26	7.70	7.83	$6.8^{+1.5}_{-1.0}$	44.27	-1.83
NGC 7469	0.90	0.70	6.87	6.88	$0.76^{+0.75}_{-0.76}$	43.14	-1.86
PG 0804+762	1.39	2.00	7.74	8.34	$21.9^{+3.8}_{-4.5}$	45.75	-0.70
PG 0953+414	1.54	2.03	7.83	8.19	$15.5^{+10.8}_{-9.1}$	45.81	-0.49

changes from a "dark halo" at the extended narrow-line region distances to being dominated by the central mass concentration in the NLR, with an almost Keplerian falloff in the $1'' < R < 4''$ interval. The observed velocity of the gas at $0\rlap{.}''5$ implies a mass of $M \sim 10^9\,M_\odot$ within the inner 60 pc. The presence of a turnover in the rotation curve indicates that this central mass concentration is extended. The first measured velocity point (outside the region saturated by the nucleus) would imply an enclosed mass of $\sim 5 \times 10^7\,M_\odot$ within $R \sim 0\rlap{.}''15$ (10 pc), which represents an upper limit to any nuclear point mass.

3 Extended Nuclear Disks

Observations of a number of extended (a few 100 pc) nuclear disks with the *HST* has provided new evidence and constraints on the mass of the MDOs in early type galaxies.

Ferrarese & Ford [18] carried out *HST* imaging and spectroscopy of NGC 6251, a giant E2 galaxy and powerful radio source which is at a distance of $\sim$ 106 Mpc. The *WFPC2* images show a well defined dust disk, 730 pc in diameter, whose normal is inclined by 76° to the line of sight. The *FOS* 0$''$09 square aperture was used to map the velocity of the gas in the central 0$''$2, from the kinematics of the gas they derive a value for the central mass concentration, 4×10^8 to 8×10^8 M$_\odot$.

Other galaxies studied with *HST* at high spatial resolution include NGC 4261, NGC 4374, NGC 7052 [19–21] and show black hole masses in the range 2–6 $\times 10^8$ M$_\odot$.

3.1 Cen A

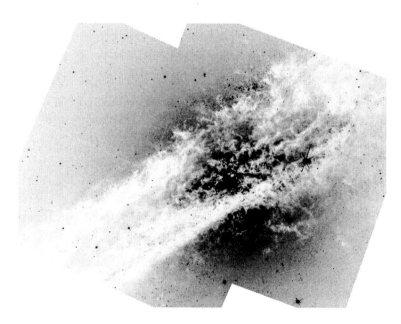

Fig. 1. Grayscale representation of the mosaic in the *WFPC2* F814W filter. Surface brightness ranges from 0 (white) to 1.6 in units of 10^{-16} erg s^{-1} cm^{-2} Å^{-1} arcsec^{-2}. Image sizes are 225$''$ × 170$''$. North is up and East is left [22]

Centaurus A (NGC 5128) is the closest (3.5 Mpc) giant elliptical galaxy hosting an active galactic nucleus (AGN) and a jet (Fig. 1). The prominent dust lane, which obscures the inner half kiloparsec of the galaxy, with associated gas, young stars and HII regions, is interpreted as the result of a relatively recent merger event between a giant elliptical galaxy and a small, gas rich, disk galaxy [23–25].

IR and CO observations of the dust lane have been modeled by a thin warped disk [26,27] which dominates ground-based near-IR observations along with the extended galaxy emission [28]. Earlier R-band imaging polarimetry from *HST* with *WFPC* [29] are also consistent with dichroic polarization from such a disk.

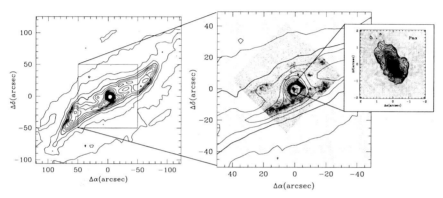

Fig. 2. (Left panel): contours from the ISOCAM image at 7 μm. **(Center panel)**: overlay of the ISOCAM contours on the NIC3 Paα image showing the morphological association between the Paα emission and the edges of the putative bar. **(Right panel)**: the Paα disk from Paper II. Note that its major axis is perpendicular to the edges of the "bar" [22]

Recent *HST WFC2* and *NICMOS* observations of Centaurus A have shown that the 20 pc-scale nuclear disk previously detected by NICMOS in Paα [30] has also been detected in the [FeII]λ1.64 μm line which shows a morphology similar to that observed in Paα with an [FeII]/Paα ration typical of low ionization Seyfert galaxies and LINERSs (Fig. 2). Marconi et al. [22] derive a map of dust extinction, E(B–V), in a $20'' \times 20''$ circumnuclear region and reveal a several arcsecond long dust feature near to but just below the nucleus, oriented in a direction transverse to the large dust lane. This structure may be related to the bar observed with ISO and SCUBA, as reported by Mirabel et al. [31]. They find rows of Paα emission knots along the top and bottom edges of the bar, with they interpret as star formation regions, possibly caused by shocks driven into the gas. The inferred star formation rates are moderately high ($\sim 0.3\,M_\odot\,yr^{-1}$). If the bar represents a mechanism for transferring gas in to the center of the galaxy, then the large dust lane across the galaxy, the bar, the knots, and the inner Paα disk all represent aspects of the feeding of the AGN. Gas and dust are supplied by a recent galaxy merger; a several arcminute-scale bar allows the dissipation of angular momentum and infall of gas toward the center of the galaxy; subsequent shocks trigger star formation; and the gas eventually accretes onto the AGN via the 20 pc disk.

By reconstructing the radial light profile of the galaxy to within $0''.1$ of the nucleus Marconi et al. [22] show that Centaurus A has a core profile. Using the models of van der Marel [32], they estimate a black hole mass of $\sim 10^9$ M$_\odot$, consistent with ground based kinematical measurements [33].

4 Statistical Properties of AGN and Radio Galaxies

An important question about AGN hosts is whether there is anything unusual about their morphology, whether they occur only in a certain type of galaxy, or can be found in all galaxies but their active phase lasts for only a fraction of a Hubble time. Furthermore, the morphology of the host may provide important information about the dynamics that funnel accretion fuel into the nucleus. Studies of the environments of quasars can also provide insight into the AGN phenomenon in general, such as the relationship between quasars and radio galaxies. If indeed quasars and radio galaxies are objects differentiated only by viewing angle, then quasars might also be expected to exhibit an alignment effect over the same redshift range as the radio galaxies.

The original classification of radio galaxies by Fanaroff & Riley [34] is based on a morphological criterion, i.e. edge darkened (FR I) vs. edge brightened (FR II) radio structure. It was later discovered that this dichotomy corresponds to a continuous transition in total radio luminosity (at 178 MHz) which formally occurs at $L_{178} = 2 \times 10^{26}$ W Hz^{-1}.

4.1 Host Morphology of Radio Galaxies and Quasars

HST observations of 273 sources in the 3CR catalog were carried out by Martel et al. [35,36]. To study the morphology distribution of the radio galaxies in the sample, they selected those at relatively small redshift. This ensures adequate image quality to permit reliable determination of the morphology, and minimizes the effects due to cosmological evolution of either the population of radio galaxies or the nature of their hosts. The result of this study is that more than 80% of the radio sources are found in elliptical galaxies, and the remainder have hosts whose morphologies are difficult to determine.

The 3CR sample is particularly well suited for investigating the relationship between radio galaxies and quasars, and the results have been discussed by Martel et al. [35,36] and Lehnert et al. [37] (Figs. 3 and 4).

The study shows that the quasar "fuzz" contributes from <5% to as much as 100% of the total light from the quasar, with a typical value of about 20%. Most of the sources are resolved and show complex morphology with twisted, asymmetric, and/or distorted isophotes and irregular extensions. In almost

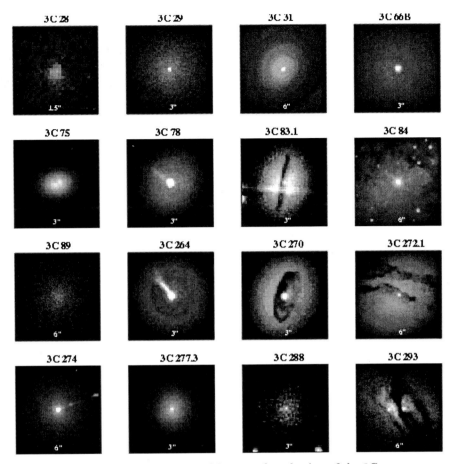

Fig. 3. *HST/WFPC2* broad band images of a selection of the 3C sources

every case of the quasars with spatially resolved "fuzz," there are similarities between the radio and optical morphologies. A significant fraction ($\sim 25\%$) of the sources show nearby galaxies in projection and $\sim 10\%$ of the sources show obvious signs of interactions with these nearby companions. These results show that the generally complex morphologies of host galaxies of quasars are influenced by the radio emitting plasma and by the presence of nearby companions.

Bahcall et al. [38] have studied in detail nine radio-loud quasars and found that the hosts are either bright ellipticals or occur in interacting systems (Fig. 5). There is a strong correlation between the radio emission of the quasar and the luminosity of the host galaxy; the radio-loud quasars reside in galaxies that are on average about 1 mag brighter than hosts of the radio-quiet quasars.

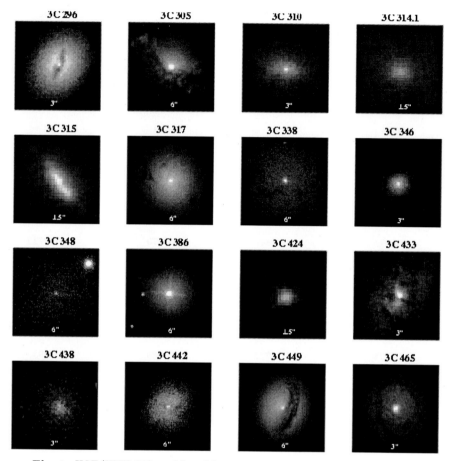

Fig. 4. *HST/WFPC2* broad band images of a selection of the 3C sources

Further *HST* observations of radio-loud quasars by Lehnert et al. [37], analyzed the spatially-resolved structures around five high-redshift radio-loud quasars.

Comparing the images with high resolution *VLA* radio images they conclude that all of the high redshift quasars are extended in both the rest-frame UV continuum and in Lyα.

The typical integrated magnitude of the host is $V \sim 22 \pm 0.5$, the typical UV luminosity is $\sim 10^{10}$ L$_\odot$, and the Lyα images are also spatially-resolved. The typical luminosity of the extended Lyα is about few $\times 10^{44}$ ergs s^{-1}; these luminosities require roughly a few percent of the total ionizing radiation of the quasar.

These results show that the generally complex morphologies of host galaxies are influenced by the radio emitting plasma. This manifests itself in the "alignment" between the radio, Lyα, and UV continuum emission, in de-

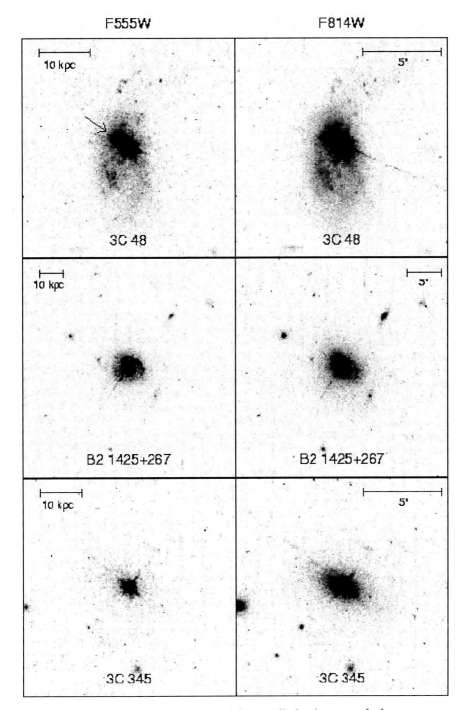

Fig. 5. The host galaxies of three radio loud quasars [38]

tailed morphological correspondence in some of the sources which suggests "jet-cloud" interactions, and in the fact that the brightest radio emission and the side of the radio emission with the shortest projected distance from the nucleus occurs on the same side of the quasar nucleus as the brightest, most significant Lyα emission.

There are few studies of radio-quiet quasars. Some early *HST* observations by Bahcall et al. [38,39] and more detailed observations by Disney et al. [40], Boyce et al. [41,42], showed that for a total of about 25 objects the parent galaxies can be either ellipticals or spirals. Overall there are six clear examples of strong ongoing gravitational interaction between two or more galaxies and in 19 other cases close companion objects are detected, suggesting recent gravitational interaction.

4.2 Seyfert Morphologies

The study of fueling processes in AGNs is key to our understanding of the structure and evolution of the central black hole and their host galaxies. Although fuel is readily available in the disk, it needs to overcome the centrifugal barrier to reach the innermost regions in disk and elliptical galaxies. Large-scale non-axisymmetries, such as galactic bars, are thought to be related to starburst activity within the central kpc, which preferentially occurs in barred hosts (e.g. [43–46]). In a number of early optical surveys, the fueling of Seyfert activity in disk galaxies was linked to non-axisymmetric distortions of galactic gravitational potentials by large-scale stellar bars and tidal interactions [47–49]. This was supported by the argument that gravitational torques are able to remove the excess angular momentum from gas, which falls inwards, giving rise to different types of activity at the center [50,51]. However early studies were not successful in showing significantly higher fractions of bars in host galaxies of AGN. Recently Knapen, Shlosman & Peletier [52] have carried out NIR observations at high spatial resolution, on a sample of 34 non-active galaxies from the CF3 catalogue as well as a sample of 48 AGN from the CFA survey. They find that Seyfert hosts are barred more often than normal galaxies, 79% ± 7.5% barred for the Seyferts, vs. 59% ± 9% for the control sample, which is 2.5σ result.

Their result suggests, but does not prove, that there is an underlying morphological difference between Seyfert and non-Seyfert galaxies, and emphasize the prevalence of barred morphologies in disk galaxies in general, and in active galaxies in particular.

4.3 Seyfert Nuclei

In the standard paradigm where AGNs are powered by non-spherical accretion onto massive black holes, the AGN's luminosity is proportional to the black hole mass accretion rate, which is about $0.01\,M_\odot$ year^{-1} for a bright Seyfert nucleus. Strong interactions or mergers with another galaxy are very efficient at funneling large amounts of gas by distorting the galactic potential and disturbing the orbits of gas clouds [53–55]. This fuel is then brought down to several thousand Schwarzschild radii, or 10^{17} cm for a black hole mass of $10^8 \, M_\odot$, at which point viscous processes drive the final accretion onto the black hole.

However, direct observational evidence that galaxy encounters stimulate the luminosity of an AGN has been ambiguous [47,49,56–60].

Malkan, Gorjiam & Tam [61] have recently published the results of an *HST* snapshot imaging survey of 256 cores of active galaxies selected from the "Catalog of Quasars and Active Nuclei" by Véron-Cetty & Véron [62,63]. Of these, 91 are galaxies with nuclear optical spectra classified as "Seyfert 1," 114 galaxies are classified as "Seyfert 2," and 51 galaxies are classified as "HIIs." This large sample of high-resolution images was used to search for statistical differences in their morphologies.

The Seyfert galaxies do not, on average, resemble the HII galaxies, which have more irregularity and clumpiness associated with their high rates of current star formation. Conversely, none of the HII galaxies have the filaments or wisps photoionized by the active nucleus which are seen in Seyfert 1 and 2 galaxies. Of the Seyfert 1 galaxies, 63% have an unresolved nucleus, 50% of which are saturated, and 6% have such dominant nuclei that they would appear as "naked quasars" at higher redshifts. The presence of an unresolved and/or saturated nucleus is anti-correlated with an intermediate spectroscopic classification (such as Seyfert 1.8 or 1.9) and implies that those Seyfert 1s with weak nuclei in the *HST* images are extinguished and reddened by dust.

The vast majority of the Seyfert 2 galaxies show no central point source. If all Seyfert 2s were to have unresolved continuum sources like those in Seyfert 1s, they would be at least an order of magnitude fainter. In those galaxies without any detectable central point source (37% of the Seyfert 1s; 98% of the Seyfert 2s, and 100% of the HIIs), the central surface brightnesses are statistically similar to those observed in the bulges of normal galaxies.

Seyfert 1s and 2s both show circumnuclear rings in about 10% of the galaxies. Malkan et al. [61] identified strong inner bars as often in Seyfert 1 galaxies (27%) as in Seyfert 2 galaxies (22%).

The Seyfert 2 galaxies are more likely than Seyfert 1s to show irregular or disturbed dust absorption in their centers as well as galactic dust lanes which pass very near their nuclei, and on average, tend to have latter morphological types than the Seyfert 1s. Thus it appears that the host galaxies of Seyfert 1 and 2 nuclei are *not* intrinsically identical. A galaxy with more nuclear dust and in particular more irregularly distributed dust is more likely to harbor a Seyfert 2 nucleus. This indicates that the higher dust-covering fractions in Seyfert 2s are the reason for their spectroscopic classification: their compact Seyfert 1 nucleus may have been obscured by galactic dust. This statistical result does not agree with the unified scheme for Seyfert galaxies, thus Malkan et al. [61] propose that the obscuration which converts an intrinsic Seyfert 1 nucleus into an apparent Seyfert 2 occurs in the host galaxy hundred of parsecs from the nucleus. If so, this obscuration may have no relation to a hypothetical dust torus surrounding the central black hole.

4.4 Morphologies of FR I Radio Galaxies

Significant progress in the understanding of the inner structure of FR I have been obtained thanks to *HST* observations. A newly discovered feature in FR I are faint, nuclear optical components, which might represent the elusive emission associated with the AGN. Their study can be a powerful tool to directly compare the nuclear properties of FR I with those of other AGNs, such as BL Lac objects and powerful radio galaxies.

Chiaberge, Capetti & Celotti [64] have studied a complete sample of 33 FR I sources from the 3CR observations carried out as part of the *HST* snapshot survey and discussed by Martel et al. [35,36] (objects with $z < 0.1$) and by de Koff et al. [65] (objects with $0.1 < z < 0.5$). Chiaberge et al. [64] have shown that an unresolved nuclear source (Central Compact Core, CCC) is present in the great majority of these objects. The CCC emission, found to be strongly connected with the radio core emission, is anisotropic and can be identified with optical synchrotron radiation produced in the inner regions by a relativistic jet. These results are qualitatively consistent with the unifying model in which FR I radio galaxies are misoriented BL Lac objects. However, the analysis of objects with a total radio power of $< 2 \times 10^{26}\,\mathrm{W\,Hz^{-1}}$, shows that a CCC is found in all galaxies except three, for which absorption from extended dust structures clearly plays a role. This result casts serious doubts on the presence of obscuring thick tori in FR I as a whole.

The CCC luminosity represents a firm upper limit to any thermal component, and implies an optical luminosity of only $\lesssim 10^{-5}$–10^{-7} times Eddington (for a $10^9\,M_\odot$ black hole). This limit on the radiative output of accreting matter is independent from but consistent with those inferred from X-ray observations for large elliptical galaxies, thus suggesting that accretion might take place in a low efficiency radiative regime [66].

The picture which emerges is that the innermost structure of FR I radio galaxies differs in many crucial aspects from that of the other classes of AGN; they lack the substantial BLR, tori and thermal disk emission, which are usually associated with active nuclei. Similar studies of higher luminosity radio galaxies will be clearly crucial to determine if either a continuity between low and high luminosity sources exists or, alternatively, they represent substantially different manifestations of the accretion process onto a supermassive black hole.

5 Demographics of Massive Black Holes

5.1 Ellipticals and S0 Galaxies

The evidence that massive dark objects (MDOs) are present in the centers of nearby galaxies has been reviewed by Kormendy & Richstone [67], Bender, Kormendy, & Dehnen [68], Kormendy et al. [69], and van der Marel [7]. The MDOs are probably black holes, since star clusters of the required mass and size are difficult to construct and maintain, and since black hole quasar remnants are expected to be common in galaxy centers. Kormendy & Richstone [67], Gebhardt et al. [4,5], and Merrit & Ferrarese [6] suggest that at least 20% of nearby kinematically hot galaxies (ellipticals and spiral bulges) have MDOs and show a correlation $M_\bullet \simeq 0.003 M_{\text{bulge}}$, where M_{bulge} is the mass of the hot stellar component of the galaxy. For a "bulge" with constant mass-to-light ratio Υ and luminosity L, $M_{\text{bulge}} \equiv \Upsilon L$.

To further probe this correlation Gebhardt et al. [4,5] constructed dynamical models for a sample of nearby galaxies with *HST* photometry and ground-based kinematics. The models assume that each galaxy is axisymmetric, with a two-integral distribution function, arbitrary inclination angle, a position-independent stellar mass-to-light ratio Υ, and a central massive dark object (MDO) of arbitrary mass $M_\bullet$. They provide acceptable fits to 32 of the galaxies, and the mass-to-light ratios inferred show a correlation $\Upsilon \propto L^{0.2}$ (Fig. 6).

The result is that virtually every hot galaxy hosts a MDO with a mass ranging from $\sim 10^8 \, M_\odot$ to $2 \times 10^{10} \, M_\odot$ and roughly proportional to the mass of the spheroidal stellar component $M_\bullet \sim 0.006 M_{\text{bulge}}$. MDO masses are just large enough to match those related to the QSO phenomenon. In fact, the highest bolometric luminosities of Quasars ($L_{\text{bol}} \lesssim 4 \times 10^{48}$ erg/s) imply, under the assumption that they radiate at the Eddington limit, that the underlying black hole masses are comparable with those of the biggest MDOs detected in ellipticals, while the lowest QSO bolometric luminosities $L_{\text{bol}} = 10^{46}$ erg/s still imply quite conspicuous black hole masses: $M_{\text{BH}} > 2 \times 10^8 \, M_\odot$.

5.2 Early and Late Type Spirals

Salucci et al. [70] have studied the rotation curves of about one thousand spiral galaxies to investigate whether they could host *relic* black holes. The sample comprises late type spirals with at least one velocity measurement inside 250 pc for 435 objects and inside 350 pc for the remaining ~ 500 objects. This would allow detections of MDOs of mass $M_{\text{MDO}} \gtrsim (1-2) \times 10^8 \, M_\odot$, typical of a black hole powering a QSO and much larger than the ordinary stellar component inside this radius.

The upper limits obtained are shown in Fig. 7: the central objects in spirals are remarkably less massive than those detected in ellipticals: strict

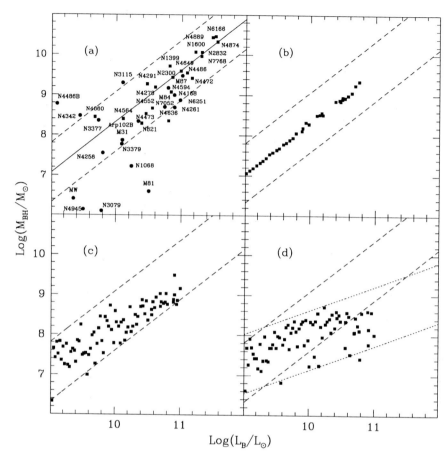

Fig. 6. (a) Observed black hole masses and bulge luminosities for the samples of nearby galaxies compiled by Ho 1998, (*circles*) and Magorrian et al. [2], (*squares*). The solid line shows the linear least square fit to the data, $\log(M_{BH} M_\odot) = 1.28 \log(L_B/L_\odot) - 4.46$, while the dashed lines show the $\pm 1\sigma$ deviation ($\sigma = 0.74$). The dashed lines are the same in all panels for comparison. (b) Monte Carlo simulations of black hole masses and bulge luminosities with $M_{acc} = 6 \times 10^{-3} \Delta M_*$. (c) Same as (b) with $M_{acc} = 1.4 \times 10^{-3}(1+z)^2 \Delta M_*$. (d) Same as (b) with $M_{acc} = 10^{-6}(1+z)^2 M_{halo} \exp[-v_c/300 \,\mathrm{km\,s^{-1}})^4]$. The linear least square fit to the results of model (iii) has a slope of 0.6, shown by the two dotted lines [4,5]

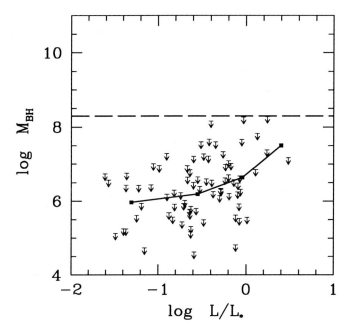

Fig. 7. The upper limits on the MDO mass as a function of luminosity. The solid line represents the corresponding luminosity-averaged values. Also shown the *minimum* mass for QSO remnants (dashed line) [71]

upper limits to their mass range are between $10^6 \, M_\odot$ at $L_B \simeq 1/20 L_*$ to $\simeq 10^7 \, M_\odot$ at $\sim 3 L_*$.

Salucci et al. [70] also analyzed Sa galaxies which are considerably fewer than late type spirals, but in view of their very massive bulge, $M_b > 10^{10} \, M_\odot$, they may well be the location of black holes with $M_{BH} \gtrsim 10^8 \, M_\odot$.

They derived MDO upper limits which range from $5 \times 10^6 \, M_\odot < M^u_{MDO} < 10^{10} \, M_\odot$, and are, at a given luminosity, one order of magnitude larger than the upper limits of the late type spirals MDOs. This implies that inside the innermost kpc, early type spirals have a large enough mass to envelop a large MDO and thus comfortably hide the remnant of a bright quasar.

6 Formation and Evolution

There is a strong empirical relationship between black holes and their host galaxies, it is therefore important to compare their properties and distribution during the quasar era at $z \sim 3$. Were today's MBHs already fully formed by that time, or was the average MBH smaller in the past and later grew through accretion or mergers to form the present population?

The epoch of maximal activity in the Universe peaked just before the epoch of maximal star formation, and MBHs must have been formed and active before this time to provide the energy to power the quasars. While the rise of luminous quasars follows closely the rise in starbirth, the bright quasars reach their peak at $z \gtrsim 2 (t \lesssim 1.6 \times 10^9 \, \mathrm{yr})$, and then their number decline about 10^9 yr before the peak in star formation, which occurs at $z \sim 1.2 (t = 2.6 \times 10^9 \, \mathrm{yr})$.

This scenario favors models in which the black hole forms before the formation of the densest parts of galaxies. For this reason we can associate the birth of quasars with the spheroid formation, a process which is closely coupled to dense regions that collapse early.

The decline in the number of quasars at $z < 2$ may be caused by several mechanisms including the exhaustion of the available fuel. Galaxy mergers, which are an effective gas transport process, became less frequent as time evolves and involve a lower mean density.

Limits to the total mass of the black hole can occur through different mechanisms. Sellwood & Moore [71] have suggested that strong bars form in the centers of recently formed galaxies and channel mass inwards to the central black hole which grows until its mass is ~ 0.02 of the mass of the disk. The bar then weakens and infalling mass forms a much more massive bulge which creates an inner Lindblad resonance which suppresses re-formation of a bar. Another mechanism proposed by Merritt [10] is that the black hole makes the central stellar orbits become chaotic, with the consequence that non-axisymmetric disturbances are smoothed out and the rate of infall of accreting gas falls.

Cattaneo et al. [72] have studied a very simple model in which both spheroids and supermassive black holes form through mergers of galaxies of comparable masses. They assumed that cooling only forms disk galaxies and that, whenever two galaxies of comparable masses merge, the merging remnant is an elliptical galaxy, a burst of star formation takes place and a fraction of the gas in the merging remnant is accreted by a central supermassive black hole formed by the coalescence of the central black holes in the merging galaxies.

This simple model is consistent with the shape of the quasar luminosity function, but its redshift evolution cannot be explained purely in terms of a decrease in the merging rate and of a decline in the amount of fuel available. To explain the evolution of the space density of bright quasars in the interval $0 < z < 2$, additional assumptions are needed, such as a redshift dependence of the fraction of available gas accreted or of the accretion time-scale.

In another scenario, proposed by Silk & Rees [73] and by Haehnelt et al. [74], a $\sim 10^6 \, M_\odot$ black hole forms by coherent collapse in the nucleus before most of the bulge gas turns into stars. If the black hole accretes and radiates at the Eddington limit, it can drive a wind with kinetic luminosity ~ 0.1 of the radiative luminosity. This deposits energy into the bulge gas, and will unbind it on a dynamical timescale with the result that the black hole mass will be limited to a value where it is able to shut off its own fuel supply [9].

A bright AGN may also limit infall of gas to form a disk, through Compton heating, radiation pressure on dust or direct interaction with a powerful wind. When the black hole mass and luminosity are large, the weakly bound, infalling gas will be blown away and an elliptical galaxy will be left behind. Only when the black hole mass is small, will a prominent disk develop. In this case, the bulge to disk ratio should correlate with the black hole mass fraction.

In summary, it seems very likely that black holes form first at quite large redshifts, $z \gg 2$ and can grow to their present sizes with standard radiative efficiency, by the time of the main quasar epoch at $t \sim 3$ Gyr. There are several plausible mechanisms to limit the growth of the black hole by switching off the fuel supply, all of which need to be need to be studied further.

Acknowledgement

It has been an honour and a pleasure to know Lo Woltjer for so many years. He has been a role model to me and to many others.

References

1. Ho, L. C. (1998). In: S. K. Chakrabarti (Ed.) Observational Evidence for Black Holes in the Universe. Kluwer, 157
2. Magorrian, J., Tremaine, S., Richstone, D., Bender, R., Bower, G., Dressler, A., Faber, S. M., Gebhardt, K., Green, R., Grillmair, C., Kormendy, J., Lauer, T. R. (1998) AJ **115**, 2285
3. Richstone, D., Ajhar, E. A., Bender, R., Bower, G., Dressler, A., Faber, S. M., Filippenko, A. V., Gebhardt, K., Green, R., Ho, L. C., Kormendy, J., Lauer, T. R., Magorrian, J., & Tremaine, S. (1998) Nature **395**, 14
4. Gebhardt, K., et al. (2000a) ApJ **539**, L9
5. Gebhardt, K., et al. (2000b) ApJ, in press, astro-ph/0007123
6. Merrit, D. & Ferrarese, L. (2000) astro-ph/0008310
7. van der Marel, R. P. (1999) AJ **117**, 744
8. Chokshi, A. & Turner, E. L. (1992) MNRAS **259**, 421
9. Blandford, R. D. (1999) In: D. Merritt, M. Jalluri, & J. A. Sellwood (Eds.) Galaxy Dynamics. ASP Conf. Series 182
10. Merritt, D. (1998) Comm. Ap. **19**, 1
11. Miyoshi, M., et al. (1995) Nature **373**, 127
12. Ichimaru, S. (1987) ApJ **214**, 840
13. Narayan, R. & Yi, I. (1994) ApJ **428**, L13
14. Abramowicz, M., Chen, X., Kato, S., Lasota, J. P., & Regev, O. (1995) ApJ **438**, L37
15. Macchetto, F. D., Marconi, A., Axon, D. J., Capetti, A., Sparks, W. B., & Crane, P. (1997) ApJ **489**, 579
16. Wandel, A., Peterson, B. N., & Malkan, M. A. (1999) ApJ **526**, 579
17. Winge, C., Axon, D. J., Macchetto, F. D., Capetti, A., & Marconi, A. (1999) ApJ **519**, 134

18. Ferrarese, L. & Ford, H. C. (1999) ApJ, in press
19. Ferrarese, L., Ford, H., & Jaffe, W. (1996) ApJ **470**, 444
20. Bower, G. A., Green, R. F., Danks, A., et al. (1998) ApJ **492**, L111
21. van der Marel, R. P. & Van den Bosch, F. C. (1998) AJ **116**, 2220
22. Marconi, A., Schreier, E. J., Koekemoer, A., Capetti, A., Axon, D. J., Macchetto, F. D., & Caon, N. (2000) ApJ **528**, 276
23. Baade, W. & Minkowski, R. (1954) ApJ **119**, 215
24. Graham, J. A. (1979) ApJ **232**, 60
25. Malin, D. F., Quinn, P. J., Graham, J. A. (1983) ApJ **272**, L5
26. Quillen, A. C., de Zeeuw, P.T., Philley, E. S., Phillips, T. G. (1992) ApJ **391**, 121
27. Quillen, A. C., Graham, J. R., & Frogel, J. A. (1993) ApJ **412**, 550
28. Packham, C., Hough, J. H., Young, S., et al. (1996) MNRAS **278**, 406
29. Schreier, E. J., Capetti, A., Macchetto, F. D., Sparks, W. B., Ford, H. J. (1996) ApJ **459**, 535
30. Schreier, E. J., Marconi, A., Axon, D. J., Caon, N., Macchetto, F. D., Capetti, A., Hough, J. H., Young, S., & Packham, C. (1998) ApJ **499**, L143
31. Mirabel, I. F., Laurent, O., Sanders, D. B., et al. (1999) A&A **341**, 667
32. van der Marel, R. P. (1999) AJ **117**, 744
33. Israel, F. P. (1998) A&A Rev. **8**, 237
34. Fanaroff, B. L. & Riley, J. M. (1974) MNRAS **167**, 31
35. Martel, A. R., Baum, S. A., Sparks, W. B., et al. (1997) BAAS **192**, 5204
36. Martel, A. R., Baum, S. A., Sparks, W. B., Wyckoff, E., Biretta, J. A., Golombek, D., Macchetto, F. D., McCarthy, P. J., de Koff, S., & Miley, G. K. (1999) ApJS **122**, 81
37. Lehnert, M. D., Miley, G. K., Sparks, W. B., Baum, S. A., Biretta, J., Golombek, D., de Koff, S., Macchetto, F. D., & McCarthy, P. J. (1999) ApJS **123**, 351
38. Bahcall, J. N., Kirhakos, S., Saxe, D. H, & Schneider, D. P. (1997) ApJ **479**, 642
39. Bahcall, J. N., Kirhakos, S., & Schneider, D. P. (1996). In: D. Clements and I. Perez-Fournon (Eds.) Quasar Hosts. Springer-Verlag, 37
40. Disney, M. J., Boyce, P. J., Blades, J. C., Boksenberg, A., Crane, P., Deharveng, J. M., Macchetto, F. D., Mackay, C. D., Sparks, W. B., & Phillips, S. (1995) Nature **376**, 150
41. Boyce, P. J., Disney, M. J., Blades, J. C., Boksenberg, A., Crane, P., Deharveng, J. M., Macchetto, F. D., Mackay, C. D., & Sparks, W. B. (1996) ApJ **473**, 760
42. Boyce, P. J., Disney, M. J., Blades, J. C., Boksenberg, A., Crane, P., Deharveng, J. M., Macchetto, F. D., Mackay, C. D., & Sparks, W. B. (1998) MNRAS **298**, 121
43. Heckman, T. (1980) A&A **88**, 365
44. Balzano, V. A. (1983) ApJ **268**, 602
45. Devereux, N. A. (1987) ApJ **323**, 91
46. Kennicutt, R. C. (1994). In: I. Shlosman (Ed.) Mass-Transfer Induced Activity in Galaxies. Cambridge Univ. Press, 131
47. Adams, T. F. (1977) ApJS **33**, 19
48. Simkin, S. M., Su, H. J., & Schwarz, M. P. (1980) ApJ **237**, 404
49. Dahari, O. (1985a) AJ **90**, 1772
50. Sellwood, J. A. & Wilkinson, A. (1993) Rep. Prog. Phys. **56**, 173
51. Phinney, E. S. P. (1994). In: I. Shlosman (Ed.) Mass-Transfer Induced Activity in Galaxies. Cambridge Univ. Press, 1

52. Knapen, J. H., Schlosman, I., & Peletier, R. F. (2000) ApJ **529**, 93
53. Shlosman, I., Frank, J., & Begelman, M. C. (1989) Nature **338**, 45
54. Shlosman, I., Begelman, M. C., & Frank, J. (1990) Nature **345**, 679
55. Hernquist, L. & Mihos, J. C. (1995) ApJ **448**, 41
56. Petrosian, A. R. (1983) Astrofizika **18**, 548
57. Kennicutt, R. C. Jr. & Keel, W. C. (1984) ApJ **2791**, 5
58. Dahari, O. (1985b) ApJS **57**, 643
59. Bushouse, H. A. (1986) AJ **91**, 255
60. Fuentes-Williams, T. & Stocke, J. (1988) AJ **96**, 1235
61. Malkan, M. A., Gorjian, V., & Tam, R. (1998) ApJS **117**, 25
62. Véron-Cetty, M. P. & Véron, P. (1986) A&AS **66**, 335
63. Véron-Cetty, M. P. & Véron, P. (1987) ESO Sci. Rep. **No. 5**
64. Chiaberge, M., Capetti, A., & Celotti, A. (1999) A&A **349**, 77
65. de Koff, S., Baum, S. A., Sparks, W. B., et al. (1996) ApJS **107**, 621
66. Fabian, A. C. & Rees, M. J. (1995) MNRAS **277**, L55
67. Kormendy, J. & Richstone, D. (1995) ARA&A **33**, 581
68. Bender, R., Kormendy, J., Dehnen, W. (1996) ApJ **464**, L123
69. Kormendy, J., et al. (1997) ApJ **482**, L139
70. Salucci, P., Ratnam, C., Monaco, P., & Danese, L. (1999) MNRAS **307**, 637
71. Sellwood, J. & Moore, E. M. (1999) ApJ, in press
72. Cattaneo, A., Haenelt, M. G., & Rees, M. J. (1999) MNRAS **308**, 77
73. Silk, J. I. & Rees, M. J. (1998) A&A **331**, L1
74. Haehnelt, M., et al. (1998) MNRAS **300**, 817

Massive Black Holes in Galactic Nuclei

A. Cavaliere

Astrofisica, Dip. Fisica, Univ. Tor Vergata, Roma, I-00133, Italy

Abstract. A discussion of the quasar-galaxy connections, and a review of astrophysical models that seek to explain the optical luminosity function of the active galactic nuclei and its cosmological evolution. Two issues are stressed: how luminosity and density evolutions arise; how the bright phase of the galactic nuclei active at high z is linked with the masses of the relic black holes being measured in the nuclei of local, inactive galaxies.

1 The Quasar Phenomenon Revisited

Recent findings highlight and rekindle two issues that long since made remarkable the Quasars and the related Active Galactic Nuclei, both as individual sources and as a population.

The *population* of the optically selected objects goes through the most sharp and non-monotonic of evolutions, now discerned out to the redshift $z = 5.80$, that is, at about 0.8 Gyr from the standard Bang (Fan et al. 2000). As was already looming out in Schmidt 1970 and Osmer 1982, and then nailed down by Schmidt 1989 and Shaver et al. 1996, the space density of the bright sources appears to rise during the first Gyrs of the Universe lifetime, and to peak at $z_Q \approx 3 \pm 0.5$; then it falls down toward us by factors $10^{-2} - 10^{-3}$. A similar message came from the radio selected QSRs (see Jackson & Wall 1999), and has been echoed by the bright X-ray selected QSs (see Della Ceca et al. 1994).

On the other hand, the total number of QSOs and Seyfert nuclei together remains around several percent of the bright galaxies out to z_Q. Such numbers have long been taken (see Woltjer 1978) to indicate an active phase lasting $\Delta t \sim 10^{-1}$ Gyr but occurring in the nuclei of most bright galaxies.

As *individual* sources, the QSs emit huge powers in the continuum from IR to X-rays and beyond, commonly exceeding $L \sim 10^{45}$ erg s^{-1}, and approaching $L \sim 10^{48}$ erg s^{-1} in some objects. The time-integrated energetics is of order $E \sim 3\,10^{60}\,L_{45}\Delta t_{-1}$ erg. But small source sizes are suggested by variability, from light months in radioquiet QSOs, down to light days or hours for the lesser outputs of some Seyferts in X-rays (setting aside the more violent behavior common in the Blazar class).

Such large energies and high compactness long since pointed toward gravitational energy as the primary source (Lynden-Bell 1969). This is because an attempt to explain E in terms of the nuclear energy $E_n \lesssim 8\,10^{-3}Mc^2$ leads one to recognize that the latter is easily dominated by the gravitational energy $E_g \sim GM^2/R$ released in the process of gathering the necessary mass within radii $R \sim 10^{15}$ cm. In fact, the condition for $E \sim E_g > E_n$ reads

$$\eta_{-1}\,\Delta t_{-1}\,L_{45}/R_{15} > 1 \,, \tag{1}$$

and is easily met by many sources if an efficiency $\eta \sim 10^{-1}$ is granted to gravitational energy release and conversion into radiation. But then the masses so condensed will be around $10^8\,M_\odot$, specifically at values

$$M_8 \approx 3\,(\Delta t_{-1}\,L_{45}\,R_{15}/\eta_{-1})^{1/2} \,. \tag{2}$$

By the same token these condensations are hard to disperse again.

In fact, over the last several years massive dark objects (MDO) in the range from a few 10^6 to a few $10^9\,M_\odot$ have been detected at the center of many local bright galaxies (see Richstone et al. 1998). Such masses are currently being found (Ferrarese & Merritt 2000; Gebhardt et al. 2000) to correlate steeply and tightly with the velocity dispersion of the surrounding galaxies.

Can one simple story link the bright, compact emissions of the QSs with their cosmological evolution, and end up with massive relics in many or most large galaxies?

2 Extreme Gravity Is Not Enough

A good start is granted by the paradigm of massive BHs accreting gas at the center of galactic nuclei (see Rees 1984).

The baryons falling into these relativistically deep gravitational potential wells swirl around in an accretion disk; on transferring outward their angular momentum they spiral in to tight orbits, before plunging into the event horizon at $R_S = 3\,10^{13}\,M_8$ cm. So by the same token the source sizes are minimized and the efficiency is maximized, indeed at $\eta \lesssim R_S/4\,R \sim 10^{-1}$ in Schwarzschild geometry (and up to $\eta \approx 0.4$ for rotating BHs).

BHs also provide the long term stability of a terminal configuration; nailing down the indication from eq. (2), they keep complete records of the mass

$$M_{BH} = \int dt\, L(t)/\eta\,c^2 \tag{3}$$

accreted throughout the whole career of the galactic nuclei.

Indeed, the detections of MDOs in currently inactive galactic nuclei provide one of the best tests of the paradigm, especially in the cases where stellar dynamics or gas rotations have been resolved down to sizes incompatible with stable condensations other than BHs. The other primary evidence is provided in a number of AGNs by the profiles of X-ray lines that are broadened and skewed as required by Doppler shifts and gravitational redshift in the vicinity of the last stable orbits (see Fabian et al. 2000).

Conversely, accreting BHs induce bolometric outputs

$$L \approx \eta c^2 \Delta m / \Delta t, \qquad (4)$$

modulated (even at constant $\eta \sim 10^{-1}$) over a wide dynamic range by the mass Δm accreted over the time Δt. Early proponents of the paradigm (Zel'-dovich & Novikov 1964, Salpeter 1964) stressed two fitting scales

$$L_E \approx 10^{46} M_8 \, erg \, s^{-1} \qquad \eta t_E \approx 5 \, 10^{-2} \, \eta_{-1} \, Gyr \qquad (5)$$

for the radiation from around the BHs and for the time involved in their growth. But these scales only apply in the Eddington regime when accretion draws from an abundant mass supply in *self-limiting* conditions; these are due to the radiation pressure reacting also at close range and on short times, so the BHs grow by $\Delta m \sim M_{BH}$ over times $\Delta t \sim \eta t_E$ to yield $L \approx L_E$.

Instead, in regimes of *supply-limited* accretion (Cavaliere & Padovani 1989, Small & Blandford 1992) the emissions may easily go sub-Eddington with $L/L_E < 1$; related evidence is mounting, see Salucci et al. 1998, Wandel 1998. In the extreme, the emissions from around several giant MDOs are now bound to levels way below those expected from minimal accretion of the surrounding hot gas with the canonical efficiency $\eta \sim 10^{-1}$ (Di Matteo et al. 2000); such thoroughly quiescent BHs require quite sub-Eddington accretion to set conditions conducive not only to advection dominated flows with $\eta \ll 10^{-1}$, but also to substantial outflows rejecting most of the gas supplied.

Thus given the paradigm, the current focus is on the QS and AGN luminosities as signals of *environmental* conditions in the host galaxies and beyond. These can lead to widely different accretion rates from well under 10^{-2} up to some $10^2 \, M_\odot \, yr^{-1}$. In addition, it is the cosmological change of the environment that conceivably holds the key to the evolution.

By itself, the BH paradigm has little to say concerning the rise and fall of the bright QS population. This is because in the Eddington regime the activities of the individual QSs constitute just flashes lasting $\eta t_E < 10^{-1}$ Gyr, short compared with the evolutionary scales $t_{ev} \sim 2$ Gyrs. In turn, the latter is short compared with the span covered by the population, now really close the Universe lifetime $t_o \approx 13$ Gyr; to wit,

$$\eta t_E \ll t_{ev} \ll t_o . \qquad (6)$$

Clearly it takes a complementary agency effective on the intermediate scales t_{ev} to *coordinate* all those flashes into a coherent rise followed by a sharp and extended fall.

3 The Role of the Environment

Here enters the other paradigm, the hierarchical growth and clustering of dark matter halos where the galaxies constitute lighter baryonic cores (White &

Rees 1978). This implies substantial dynamical events that can cause large scale instabilities within the host galaxy and drive accretion at rates with a wide range and a strong epoch dependence.

In fact, mainly subgalactic or galactic halos with masses $M < 10^{13} M_\odot$ are built up through direct merging of smaller units before $z_G \simeq 2.5 \pm 0.5$ (depending on cosmogonical and cosmological details). Thereafter the galaxies begin to assemble into small groups of mass $M_G \gtrsim 10^{13} M_\odot$, where they recurrently interact with their companions and evolve or even coalesce to saturate their growth. Small groups in turn merge into richer ones and eventually into clusters. During such developments the probabilities of forming DM structures in each mass range grow rapidly at first, then go into a slow demise. But the baryons contained in the DM potential wells live more hectic times, which *amplify* their radiative behavior.

This is because large scale dynamical events like merging and interactions break the axial symmetry, or enhance the static asymmetries of the gravitational potential. Then the specific angular momentum j of the gas providing support in the central kpc of the host is not conserved, rather it is transferred to the massive DM component. This provides the necessary condition for setting the gas on an inward track; at smaller scales dissipative processes can take over in redistributing j (Haehnelt & Rees 1993), and the gas can reach the nucleus to grow new BHs or refuel the old ones.

The outputs expected from such events may be visualized in the form

$$L = \eta c^2 f m/\Delta t , \qquad (7)$$

in terms of the fraction $f \equiv \Delta m/m$ accreted out of the gas mass $m(z, M)$ residing in the host. The event probability $P(f)$ defines range and shape of the luminosity function $N(L, z)$. Their changing rate governs the source number, producing "density evolution". On average, $L \propto m(z)$ decreases and yields "luminosity evolution" when fresh gas is no longer imported into the host, while the residual supply is used up in accretion episodes and accompanying starbursts that concur with the ongoing quiescent star formation.

In treating these points, two approaches have been taken. One (Cattaneo, Haehnelt & Rees 1999, Kauffmann & Haehnelt 2000) uses semi-analytic modeling to combine in a complex numerical package two main blocks: detailed hierarchical probabilities of structure forming, provided by Monte Carlo simulations; simple phenomenological prescriptions for $f(M), \Delta t(z)$ and the efficiency of star formation, adopted as to match from the outset the correlation of BH with galaxy masses, and to accord with the main features of the QSO LF together with the history of star formation.

Alternatively, one may seek to disentangle and address directly the processes involved in nuclear fueling. To that effect, the triggering dynamical events are conveniently classified into two main regimes, with the divide provided by $z_G \approx 2.5$ to a first approximation.

4 QSs in Newly Forming Spheroids

Before z_G, successive major *merging* events build up protogalactic and galactic spheroids within similarly sized DM halos, in conditions where the baryon density $\rho \sim 10^{-1} \rho_{DM} \propto (1+z)^3$. These events strongly distort the gravitational potential causing $f \approx 1$ to hold in eq. (7), while they also import into the host much fresh gas sustaining its amount at $m(z, M) \sim 10^{-1} M$.

Then central BHs can form and/or accrete rapidly, fed by $\Delta m \sim M_{BH}$ over dynamical times t_{dyn} close to the Salpeter time; thus they shine at their full rate $L \approx L_E \propto M_{BH}$. As a result, the average luminosities of these early QSs grow and the numbers rise, tracking the formation of galactic halos as these progressively cover the range from $M \sim 10^{10}$ toward $\sim 10^{13} M_\odot$.

This picture of early QSs shining in newly forming galaxies has won wide acceptance. Direct evidence from imaging of galaxies in the process of forming around early, optically selected QSs is still meager and hard to obtain, but pioneering observations (see Fontana et al. 1998, Djorgowski 1998) have highlighted a number of good candidates.

In specific modeling, the probability of such merging events is often subsumed into the standard hierarchical distribution of the halo masses (Press & Schechter 1974). The source LF may be directly related to the latter once L is given in terms of M, on using the transformations $N(M, z)\, dM = N(M_{BH}, z)\, dM_{BH} \propto N(L, z)\, dL$. The result must be complemented with the prefactor $\eta t_E / t$ to take care of the short visibility of the sources.

For example, Haehnelt & Rees 1993 considered BH growth by coalescence in parallel with their halos merging, a process represented with the scaling

$$M_{BH} \approx 10^{-4}\, M\,. \tag{8a}$$

A prefactor $(1+z)^6$ is also expected, due to the longer time taken for two BHs to coalesce at lower densities (Monaco, Salucci & Danese 2000).

Alternatively, Haehnelt, Natarajan & Rees 1998 considered that during halo merging a central BH also accretes gas, up to the limit set by the feedback from radiative energy deposition. This can blow out much gas from the host structure, though with low (and uncertain) efficiency $f_k \sim 10^{-3}$, and the limit reads $f_k\, L_E\, t_{dyn} \lesssim G M m / r$; with the expected $m \sim 10^{-1} M$, this yields

$$M_{BH} \lesssim 10^{-3} ((1+z)/4)^{5/2} M_{13}^{2/3}\, M\,. \tag{8b}$$

These models are particularly attractive (CV98) at high z where frequent merging activity sustains the gas reservoirs; fig. 1 shows the resulting LFs. In both cases the normalization is such that the BH number matches that of actively star forming galaxies at $z \approx 3$ (Steidel et al. 1999). But the non-linear model (8b) privileges the upper range of galactic halos, where the numbers provided by the PS74 distribution are smaller, and in fact close to one BH per bright galaxy. In addition, the transformation (8b) stretches the distribution to yield flatter, more fitting optical LFs; more on this in §7.

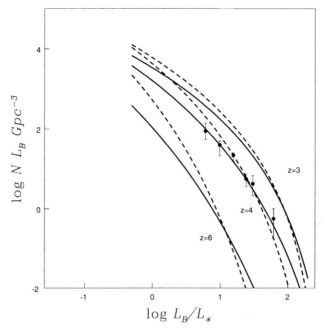

Fig. 1. High-z optical LFs from the models discussed in §4: eq. (8a), dashed; eq. (8b), solid. Here L_B is the luminosity in the blue band (bolometric correction $\kappa = 10$), and $L_* = 10^{45}$ erg s^{-1}; low density, flat cosmology with $\Omega_o = 0.3$, $\Omega_\lambda = 0.7$, $h = 0.65$. Data from Kennefick et al. 1995; Schmidt et al. 1995

5 QSs in Interacting Galaxies

Subsequent to $z \approx 2.5$ the host galaxies become enclosed in groups, where the prevailing dynamical events are best described in terms of *interactions* with a comparable or smaller companion. Some of these lead to coalescence into giant galaxies; but most do not import much fresh gas into the host, yet they again distort the potential, destabilize the residual gas and trigger *supply-limited* episodes of mass inflow and accretion. They (concurring with quiescent star formation) deplete $m(z)$ and cause the luminosities to decrease after eq. (7).

Small groups constitute preferred sites for interactions to occur and refuel old BHs, particularly in dominant galaxies. This is due to their high density of galaxies n_g, and to their low velocity dispersion V still so close to the galaxian v as to allow dynamical resonance. With nearly geometrical cross section $\Sigma \approx \pi (r_g + r_g')^2$, the expected time scale is $\tau_r \sim 1/n_g \Sigma V \sim 1$ Gyr.

Passages with impact parameters $b \sim$ a few r_g will take times b/V close to the galaxian scale $\tau = r_g/v \sim 10^{-1}$ Gyr. But gas inflow to the center takes a few additional τ's to develop, so the effective $\Delta t \sim 2\tau$. Finally, these episodes are expected to gradually dwindle and peter out as groups are reshuffled into clusters with lower n_g and larger V.

Relevant observational evidence is being provided by sharp imaging of bright QSs at intermediate z; this has elicited many host galaxies which are not really newly forming (e.g., Nolan et al. 2000) but rather are surrounded by a complex environment (see Hutchings, Crampton & Johnson 1995; Rafanelli, Violato & Baruffolo 1995; Hasinger et al. 1997; Bahcall et al. 1997; Boyce et al. 1998; McLure et al. 1999; Ridgway et al. 1999; Boyce, Disney & Bleaken 1999). Some hosts appear to harbor secondary nuclei; others are found in close association with comparable or fainter companions. A fraction $\sim 10^{-1}$ still exhibit morphological marks of a recent interaction; this should be weighted with the short duration (a few 10^{-1} Gyrs) of such features and with the time-lag of some τ's from these outer events to the central BH fueling. In addition, statistical evidence (Fisher et al. 1996; Jaeger, Fricke & Heidt 1999; Wold et al. 1999) points toward a poor group environment also for radioquiet QSOs.

That encounters prefer to occur in groups is also indicated by many N-body experiments (e.g., Governato, Tozzi & Cavaliere 1996; Athanassoula 2000). High resolution simulations including hydro treatment of the gas (see Barnes & Hernquist 1998, Mihos 1999) show the dynamical steps that follow a grazing encounter between equals. The sequence started by the gravitational torques includes the external fireworks of tidal tails and debris spread out, continues with the internal response of the host (possibly amplified by metastability conditions), and ends with $f \approx 50\%$ of the gas driven into the central 10^2 pc, the resolution limit. At the other extreme, a dwarf's interaction with a dominant galaxy may be treated in terms of dynamical friction.

Intermediate events originate from comparable companions passing by the host and easily yield $f \approx$ a few %, since in a group the effective b for two-body encounters is smaller than the group radius while $V \sim v$ holds. In a host still gas rich with $m \sim 10^{10} M_\odot$, such destabilized fractions are enough to yield outputs $L_b \sim 10^{46}$ erg s^{-1} after eq. (7). Higher L require larger f, which in turn imply closer encounters with larger companions; these events are fewer, and lead to a steepening LF.

6 Gas Supply from Interactions, a Specific Model

As the optical observations are providing highly resolved LFs (over 5000 QSOs in the range $z = 0.35 \div 2.3$ are comprised in the survey by Boyle et al. 2000), it is worth discussing in some detail the outcomes expected from the interactions. Consider specifically the direct, time-integrated action of the external torques on the host gas in equilibrium at $r \sim$ kpc; the fractional gas so made available may be written in the form (Cavaliere & Vittorini 2000)

$$f \lesssim \Delta j/j \approx r\, v\, M'/M_o\, V\, b\,, \tag{9}$$

with a postfactor r/b in the case of a truly tidal effect. Note the dependences on the orbital parameters b and the partner mass M', and on the host structural parameters, the mass $M_o(<r)$ and $v \approx (G\,M_o/r)^{1/2}$.

Toward the *shape* of the LF, consider that the hosts constitute a narrow, bright subset of galaxies (Smith et al. 1986), while the ranges of b and of V are limited as said above. So the statistics of f is dominated by the steep distribution of substantial companions ranging from $M' \sim 10^{11}\,M_\odot$ up to the host mass; for example, from the PS74 form one closely obtains $P(f) \propto f^{-2}$ before going into a cutoff. In turn, $N(L)$ is dominated by $P(f)$ since $m(z)$ is nearly constant over the encounter duration τ.

The gas inflow toward the nucleus develops over comparable times, and the sources brighten up up to the value of L allowed by the fraction accreted. All sources can attain $L \sim L_b \sim 10^{46}$ erg s^{-1} corresponding to the minimal $f \sim$ a few %; so for fainter L their number is conserved and $L\,N(L) \approx$ const applies. Those endowed with higher f will attain a proportionally higher L, then fade off and drop out of the LF following

$$NL \propto P(L \propto f) \propto L^{-2}\,, \qquad (10)$$

as shown in detail by CV00.

In sum, the LF fed by encounters is expected to *break* around 10^{46} erg s^{-1} from about $N(L) \propto L^{-1}$ to L^{-3}. Steeper slopes obtain on considering at the faint end also the sources in the process of fading out, and at the bright end torques acting indirectly and causing a weaker dependence of f on M'.

Toward the *evolution* of the LF, consider that over scales $\tau_r \gg \tau$ the gas in the host is substantially depleted by the interactions themselves, following

$$dm/dt \approx -m\,\langle f \rangle / \tau_r \qquad (11)$$

on average. Averaging with $P(f)$ yields for $\langle f \rangle$ a factor of 3 over the minimum $f \approx 1 \div 1.5\%$. But what counts for the gas depletion is the total destabilized fraction; in fact, the gas funneled inward may end up not only in accretion down to the central BH, but also in a less constrained nuclear starburst that uses at least twice as much mass (see Sanders & Mirabel 1996). This contributes another factor 3, so the true scale for gas depletion is around $6\,\tau_r$.

Finally, the latter changes with epoch as $\tau_r(z) \propto (1+z)^{-3/2}$, since in groups $\tau_r \approx 3\,t_{dyn}$ closely obtains as discussed by Cavaliere et al. 1993. The normalization to local, virialized systems may be scaled from the census of bright galaxies that show clear signs of interactions in the local field (Toomre 1977); the corresponding time of $2\,10^2$ Gyr scales with the inverse of the density to about 1 Gyr. The result is a fast depletion of the gas on the scale $\tau_r/\langle f \rangle \approx 6\,t/t_o$ Gyr in the critical universe; this gives $m(t) \propto t^{-2}$, and

$$L_b \propto m(z) \propto (1+z)^3\,. \qquad (12)$$

This strong LE is due to the average interactions becoming *ineffective* in BH feeding as the host gas reservoirs are depleted by previous discharges. A milder, DE component is due to substantial encounters becoming *infrequent* in later, richer, but less dense groups; this causes the LF amplitude to decrease like $N_G(z)/\tau_r(z)$, that is, like $(1+z)^{1.5}$ or somewhat stronger at epochs $z \lesssim 1$ when the group space density $N_G(z)$ is on its demise.

The full result from a formal calculation (CV00) is

$$N(L,z) = \tau N_G(z)/\tau_r(z) L_b \times 1/[L/L_b + (L/L_b)^3] \,, \qquad (13)$$

which turns out to be pleasingly close to the empirical formula often used to fit the optical observations, e.g., Boyle et al. 1988, La Franca & Cristiani 1997, Boyle et al. 2000. Recall that a somewhat steeper shape obtains if f depends on M' less than linearly, and that a faster gas depletion obtains from including the quiescent star formation.

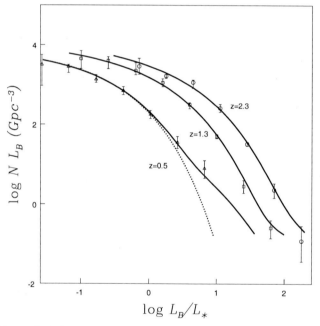

Fig. 2. The optical LF of QSOs fueled by interactions, computed from eq. (13). Cosmology, L_B, L_* as in fig. 1. Data from Boyle et al. 1988

Fig. 2 represents eq. (13) computed in the currently popular low density, flat cosmology. At the bright end two small contributions are added: (i) BHs fueled in the rare ($\tau_r \sim 10^2$ Gyr) encounters with a substantial partner of the more numerous (10 ×) hosts in the "field"; (ii) BHs flaring up to $L \sim L_E$ in the minority of galaxies formed later than z_G by some major mergers still occurring. Both these contributions are small ($\lesssim 10\%$) but decrease slowly, because newly formed galaxies import gas, and field hosts retain more gas from their history of mainly quiescent star formation (Guiderdoni et al. 1998). Thus at low z a bright excess is bound to emerge, similar to that observed by Grazian et al. 2000.

7 Relics

The mass distribution $N(M_{BH}, z)$ of the BHs formed at *high* z in growing spheroids is directly related to the QSO LF by the models (8a) and (8b), which imply the scaling relations

$$M_{BH} \propto \sigma^3 \rho^{-1/2}(z) \propto \sigma^4, \quad M_{BH} \propto \sigma^5, \qquad (14a, b)$$

respectively. These relations stem from the hierarchical scaling $R \propto (M/\rho)^{1/3}$ and $V^2 \propto GM/R$ for the surrounding DM halos as appropriate for $M < 10^{13} M_\odot$, assuming also $\sigma \propto V$ (see Kauffmann & Haehnelt 2000); the z-dependent fuzz in eq. (14a) is fixed on expressing the density at formation of the mass M with the approximate relation $\rho \propto M^{-1/2}$.

At *later* epochs $z < z_G$ the relic masses grow as $M_{BH}(z) = M_{BH}(2.5) + \sum_i \Delta m_i$, due to the additional accretions Δm_i caused by host interactions. Correspondingly, $N(M_{BH}, z)$ is remolded as given by

$$\partial N / \partial t = N_G/N\tau_r \times \int df\, P(f)\left[N(M_{BH} - fm) - N(M_{BH})\right], \qquad (15)$$

in terms of the BH number N including the dormant ones, and of the same $N_G(z)$, $\tau_r(z)$, f, $P(f)$ discussed in §6. The result shown in fig. 3 confirms that later interactions add relatively more to small than to large, early BHs.

For such large masses the relations (14a, b) generated at early z basically persist down to the relics in the local inactive galaxies, and turn out to be in tune with the current debate concerning the MDO data.

These have been recently recognized to follow a tight and steep correlation, whose precise shape is given as slightly flatter than $M_{BH} \propto \sigma^4$ by Gebhardt et al. 2000, or as nearly $M_{BH} \propto \sigma^5$ by Ferrarese & Merritt 2000. In either case the scatter is found to be small, factors $10^{\pm 0.35}$ or less in M_{BH}. In contrast, note from fig. 1 that model (8a) for the LF, differently from model (8b), does require a significant scatter ($10^{\pm 0.5}$ or more) of M_{BH} at given M to convolve the LF over (Haiman & Loeb 1998), if the flat shape observed at high z is to be fitted.

At lower z, the gas masses $\Delta m = mf \propto m(v)/v$ produced by interactions yield the fitting LF discussed in §6, and imply two features. The scatter is within an overall factor 5, the effective range of $P(f)$; note that the range $L < L_b$ does not contribute, being populated by sources still brightening before their maximum. Here the scaling reads $\Delta m \propto \sigma^3$, on approximating $v \approx \sigma$ and using the galactic luminosity-velocity relation to estimate the

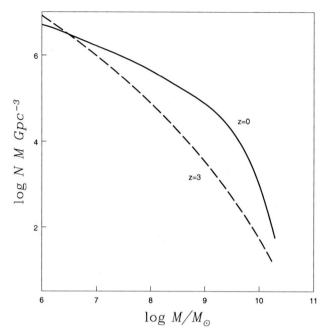

Fig. 3. The evolution of the mass distribution $N(M_{BH}, z)$ of relic BHs discussed in §7. The corresponding local mass density reads $\rho_{BH} = 3\,10^{15}\,h^3\,M_\odot\,\text{Gpc}^{-3}$

baryonic $m(\sigma) \propto \sigma^4$. Truly tidal interactions (see eq. 9) yield a steeper scaling $\Delta m \propto m(\sigma)\,r/\sigma \propto \sigma^4$ and more scatter. Such masses contributed by late accretions – if not further constrained – easily dominate for $\sigma < 150$ km s^{-1} to yield there the softer correlation $M_{BH} \propto \sigma^3 \div \sigma^4$ (see fig. 4).

8 What Next

In that critical range the MDO data are scanty and difficult to obtain in normal galaxies from disk or star dynamics; considerable extension of the reverberation mapping measurements in current AGNs is needed (see Nelson 2000). Even at larger σ the precise slope of the correlation is yet to be sorted out.

To illustrate the impact expected from these data, a steep correlation found to hold tightly for small as well as for large masses will indicate the surrounding galactic structure to exert gravitational control on the actual accretion rates all the way from early to late z.

An interesting candidate for this role (Cavaliere & Vittorini in preparation) is provided by QS feedback similar to that underlying the high-z model described by eqs. (8b), (14b). At lower z the limit reads $L \lesssim m\sigma^2/\tau\,f_k$; with $f_k \sim 10^{-3}$, this constrains the gas actually accreted to stay below what is

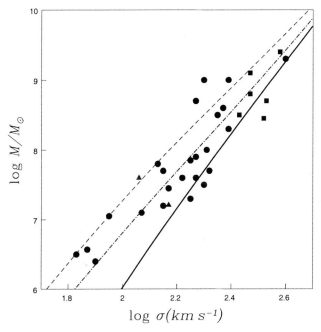

Fig. 4. The extreme models for the $M_{BH} - \sigma$ relation discussed in §6 and §7: unconstrained accretion (dot-dashed for $f = 1.5\%$, dashed for $f = \langle f \rangle = 4.5\%$); feedback-constrained accretion (solid). Data from Gebhardt et al. 2000

produced by the interactions of galaxies with $\sigma < 250$ km s^{-1}. Then the scaling $L\tau \propto m\sigma^2$ applies, and yields $\sum_i \Delta m_i \sim M_{BH} \propto \sigma^{6 \div 5}$. To stress the extremes, the steep correlation $M_{BH} \propto \sigma^5$ obtains at all σ's if QS feedback *always* provides the main constraint; if this is *never* important, the softer correlation $M_{BH} \propto \sigma^4$ holds at high σ, then goes into $M_{BH} \propto \sigma^4 \div \sigma^3$.

In this context two issues pointed out by Haehnelt at al. 1998 have still to be settled: to focus the time-scales taken by the QS feedback for reacting on the gas inflow; to bound the conceivable vagaries of the fraction f_k deposited as outflow kinetic energy, that by themselves would introduce a large scatter. Effective QS feedback may also impact the star formation history (Silk & Rees 1998), even more than the Supernovae; the data concerning QSs or AGNs and those concerning the star formation rates may be jointly used to fix the uncertainties in the process.

In conclusion, the question ending §1 requires an articulated answer. The QS story centers on *one* kind of engine (the BH) based on strong gravity; it comprises *different* regimes of fuel production, all based on gas destabilization at large scales under weak gravity, but specifically related to the environment changing from spheroids to groups; ultimately, galactic gravitation may control the throttle and *unite* to some extent the fueling modes.

The grand design looming out converges broadly with the outcomes from the other approach by Cattaneo et al. 1999 and Kauffmann & Haehnelt 2000. At early $z \gtrsim 2.5$ much fresh gas is available for full, *self-limiting* accretion; BH growth and associated QS flaring up to Eddington luminosities are geared to spheroid buildup in relatively small DM halos, governed by short dynamical scales still close to the Salpeter time. Here the LF *grows* in range and height tracking the progressive halo build up; meanwhile, a correlation is generated with a shape steep up to the form $M_{BH} \propto \sigma^5$.

Later than $z \approx 2.5$ the dearth of gas curbs the accretion feast and turns the QS rise into a descent. This is because imports of fresh gas into the hosts cease, while substantial dynamical events still involve the hosts and trigger accretion and starbursts which deplete the gas reservoirs; then *supply-limited* accretion produces fast *luminosity* evolution of the LF. Strong interactions causing bright luminosities are rare, which produces the steepening shape of the LF. The latter may be further trimmed at the bright end when the accretion is also *feedback-constrained*, which also concurs with starbursts to tighten the correlation of bright QSOs with bright/massive galaxies.

At z lower yet the gas reservoirs in the hosts are bound to approach exhaustion, and lesser gas productions or smaller supplies become relevant for weaker AGN activity; such are internal galactic instabilities (see Heller & Shlosman 1994, Merritt 1999), or gas supplied by satellite galaxies cannibalized by the hosts. The latter can fuel many weak AGNs (CV00), more likely to be pinpointed in X-rays, and mainly undergoing *density* evolution (L dependent) as the event rate drops with the depletion of the satellite retinues.

All these supply-limited accretion modes proceed on average toward sub-Eddington luminosities, except when larger gas parcels happen to overwhelm small seed BHs; as a result, L/L_E will decrease on average at fainter L, but with a large scatter. Eventually, very sub-Eddington conditions will prevail and the accretion flows ought to go into ADAF conditions with small effective $\eta \ll 10^{-1}$, a natural gateway to the debated condition of quiescent BHs.

Out of this complex story stands a direct link of the high-z activity with the massive MDOs in local, inactive galaxies. For example, *flat* optical LFs are related by the model (8b) with the *steep* correlation (14b), a link favored by the *combined* evidence as it currently stands from the *local* MDO correlation and from optical, large area surveys at *high-z* with selection down to the near IR, such as the ongoing Sloan Digital Sky Survey.

Meanwhile, enticing news is being provided by the deep surveys in X-rays from *Chandra* and *XMM-Newton*; these are resolving the dominant AGN contribution to the XRB and uncovering much hitherto hidden accretion power (cf. Hasinger, this Conference). They ought to solve long standing questions concerning any large population of optically obscured QSs at high z. But it remains to be settled how much L_X contributes to the bolometric outputs of the high-z, radioquiet QSOs and to the making of large BHs.

References

1. Athanassoula, E. 2000, Proc. IAU Symp. 174, ASP CS 209, 245
2. Bahcall, J.N., Kirhakos, S., Saxe, D.H. & Schneider, D.P. 1997, ApJ, 479, 642
3. Barnes, J.E. & Hernquist, L.E. 1998, ApJ, 495, 187
4. Boyle, B.J., Shanks, T. & Peterson, B.A. 1988, MNRAS, 235, 935
5. Boyce, P.J. et al 1998, MNRAS, 298, 121
6. Boyce, P.J., Disney, M.J. & Bleaken, D.G. 1999, MNRAS, 302, L39
7. Boyle, B.J. et al. 2000, preprint astro-ph/0005368
8. Cattaneo, A., Haehnelt, M.G. & Rees, M.J. 1999, MNRAS, 308, 77
9. Cavaliere, A., Colafrancesco, S. & Menci, N. 1993, ApJ, 392, 41
10. Cavaliere, A. & Padovani, P. 1989, ApJ, 340, L5
11. Cavaliere, A. & Vittorini, V. 1998, ASP CS, 146, 26 (CV98)
12. Cavaliere, A. & Vittorini, V. 2000, preprint astro-ph/0006194 (CV00)
13. Della Ceca, R. et al. 1994, ApJ, 430, 533
14. Di Matteo, T. et al. 2000, MNRAS, 311, 507
15. Djorgowski, G., 1998, preprint astro-ph/9805159
16. Fabian, A.C. et al. 2000, preprint astro-ph/0004366
17. Fan et al. preprint astro-ph/0005414
18. Ferrarese, L. & Merritt, D. 2000, preprint astro-ph/0006053
19. Fisher, K.B., Bahcall, J.N., Kirhakos, S. & Schneider, D.P. 1996, ApJ, 468, 469
20. Fontana, A. et al., 1998, AJ, 115, 1225
21. Gebhardt, K. et al. 2000, preprint astro-ph/0006289
22. Governato, F., Tozzi, P. & Cavaliere, A. 1996, ApJ, 458, 18
23. Grazian, A. et al. 2000, AJ, 119, 2540
24. Guiderdoni, B., Hivon, E., Bouchet, F.R., & Maffei, B. 1998, MNRAS, 295, 877
25. Haehnelt, M.G., Natarajan, P. & Rees, M.J. 1998, MNRAS, 300, 817
26. Haehnelt, M.G. & Rees, M.J. 1993, MNRAS, 263, 168
27. Hasinger, G. et al. 1997, Astron. Nachr., 6, 329
28. Haiman, Z. & Loeb, A. 1998, preprint astro-ph/9811395
29. Heller, C.H. & Shlosman, I. 1994, ApJ, 424, 84
30. Hutchings, J.B., Crampton, D. & Johnson, A. 1995, AJ, 109, 73
31. Jackson, C.A & Wall, J.V. 1999. MNRAS, 304, 160
32. Jaeger, K., Fricke K.J. & Heidt, J. 1999, preprint astro-ph/9911101
33. Kauffmann, G. & Haehnelt, M.G. 2000, MNRAS, 311, 576
34. Kennefick, J.D., Djorgowski, S.G. & de Carvalho, R.R. 1995, AJ, 110, 2553
35. La Franca, F. & Cristiani, S. 1997, AJ, 113, 1517
36. Lynden-Bell, D. 1969, Nature, 223, 690
37. McLure, R.J. et al. 1999, MNRAS, 308, 377
38. Merritt, D. 1999, Comm. Mod. Phys. 1, 39
39. Mihos, J.C. 1999 preprint astro-ph/9903115
40. Monaco, P., Salucci, P & Danese, L. 2000, MNRAS, 311, 279
41. Nelson, C.H. 2000, preprint astro-ph/0009188
42. Nolan, L.A., et al. 2000, preprint astro-ph/0002020
43. Osmer, P.S. 1982, ApJ, 253, 28
44. Press, W.H. & Schechter, P.L. 1974, ApJ, 187, 425 (PS74)
45. Rafanelli, P., Violato, M. & Baruffolo, A. 1995, AJ, 109, 1546
46. Rees, M.J. 1984, ARAA, 22, 471
47. Richstone, D. et al. 1998, Nature, 395, A14

48. Ridgway, S. et al. 1999, preprint astro-ph/9911049
49. Salucci, P., Szuszkiewicz, E, Monaco, P. & Danese, L 1998, MNRAS, 307, 637
50. Salpeter, E.E. 1964, ApJ, 140, 796
51. Sanders, D.B., & Mirabel, I.F. 1996, ARAA, 34, 749
52. Schmidt, M. 1970, ApJ 162, 371
53. Schmidt, M. 1989, Highlights Astronomy, 8, 31
54. Schmidt, M., Schneider, D.P. & Gunn, J.E. 1995, AJ, 110, 68
55. Shaver, P.A. et al. 1996, Nature, 384, 439
56. Silk, J. & Rees, M.J. 1998, A&A 331, L1
57. Small, T.A. & Blandford, R.D. 1992, MNRAS, 259, 725
58. Smith et al. 1986, ApJ, 306, 64
59. Steidel, C.S. et al. 1999, ApJ, 519, 1
60. Toomre, A. 1977, in "Galaxies and Stellar Populations", ed. B.M. Tinsley & R.B. Larson (New Haven: Yale Univ. Obs.), p. 401
61. Wandel, A. 1998, preprint astro-ph/9808171
62. White, S.D.M. & Rees, M.J. 1978, MNRAS, 183, 341
63. Wold, M., Lacy, M., Liljie, P.B. & Serjeant, S. 2000, MNRAS, 316, 267
64. Woltjer, L. 1978, Phys. Scripta, 17, 367
65. Zel'dovich, Ya.B & Novikov, I.D. 1964, Doklady Akad. Nauk SSSR 158, 311

X-Ray Studies of Active Galactic Nuclei

Yasuo Tanaka

Max-Planck-Institut für Extraterrestrische Physik, D-85748 Garching, Germany
and
Institute of Space and Astronautical Science, Sagamihara, Kanagawa-ken, Japan

Abstract. AGN are strong X-ray emitters, supermassive black holes powered by mass accretion. X-ray observations allow to probe the direct vicinity of the black holes and the circum-nuclear environment. Results of X-ray studies of active galactic nuclei are summarized on selected topics. Related subjects on luminous infrared galaxies and the cosmic X-ray background are also discussed.

1 Introduction

This paper is intended to outline what we have learned from X-ray studies of active galactic nuclei (AGN) on several selected topics.

AGN are very luminous X-ray sources. The X-ray sky is increasingly dominated by AGN as the photon energy increases. This is because the AGN have the hardest spectra among other classes of sources (νL_ν of AGN is proportional to $\sim \nu^{+0.3}$). Despite far smaller apertures of the X-ray telescopes available than big ground-based optical telescopes, similarly far or even farther AGN can be observed in X-rays. For instance, *Chandra* and *XMM-Newton* could detect QSO of $L_x \sim 10^{46}$ erg s^{-1} up to $z \sim 8$ if not confusion limited.

AGN are believed to be mass-accreting supermassive black holes. Among radiations at other wave lengths, X-rays come from the innermost region near the black holes, hence suited for studying the black holes as in the case of stellar mass black holes in galactic X-ray binaries. As we shall see, X-ray spectroscopy has become more and more important. The new-generation missions, *Chandra* and *XMM-Newton*, are capable of high-resolution spectroscopy, and will deepen our understanding of physics of AGN a great deal.

2 Accreting Supermassive Black Holes

It has been known that there are close similarities in X-ray properties between AGN and Galactic black-hole X-ray binaries. This makes it probable that the accretion processes in AGN take place in much the same way as in black-hole X-ray binaries.

2.1 X-Ray Spectrum

Black-hole X-ray binaries exhibit a bimodal behavior (see e.g. Tanaka & Shibazaki 1996). When the accretion rate is low (typically $L_X < 10^{37}$ erg s^{-1}, or below $\sim 1-3\%$ of the Eddington limit), their X-ray spectra are hard, expressed by a power law of an energy index $\alpha \sim 0.7$ with a high-energy cut-off around 100 keV. This spectrum is generally interpreted to be formed by thermal Comptonization (Sunyaev & Titarchuk 1980), but the detail is still the subject of current study. This spectral state is called the hard/low state.

On the other hand, at higher accretion rates, the spectrum changes into a characteristic form consisting of a luminous soft thermal component and a hard power-law tail. An example is shown in Fig. 1. There are enough reasons to believe that this soft thermal component is a multi-color blackbody emission expected from an optically-thick geometrically-thin accretion disk. The power-law tail is definitely steeper, with $\alpha \sim 1.5$, than the power law in the hard state, and no cut-off has been observed up to 1 MeV. This spectral state is called the soft/high state. The origin of the power-law tail is not satisfactorily understood yet.

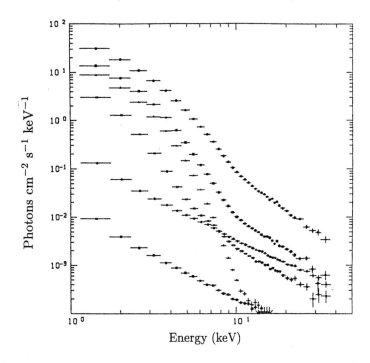

Fig. 1. Bimodal behavior of the black-hole binary GS 1124-684. The upper four spectra are in the soft state, and the lower two in the hard state

Seyfert 1s and most QSOs show an X-ray spectral shape very similar to that in the hard state of black-hole binaries, with essentially the same α, ~ 0.7 (Fig. 2), although the X-ray luminosity is orders of magnitude different between the two systems. These Seyferts and QSOs are generally considered to be in a state corresponding to the hard state of black-hole binaries, similarly at a relatively low accretion rate. Spectra of Seyfert 1s generally exhibit a high-energy cut-off also around 100 keV. For the thermal Comptonization model, the cut-off energy represents the electron temperature that is determined by the balance between heating and Compton cooling. The same α and the same cut-off energy for black-hole binaries and these AGN is remarkable, but the reason for this similarity (and why ~ 0.7?) is still unexplained.

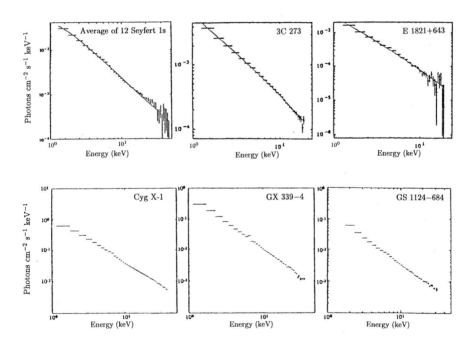

Fig. 2. Examples of the spectra of AGN (upper panel) and those of black-hole binaries in the hard state (lower panel)

Narrow-line Seyfert 1s (NLS1s) have some spectral resemblance to the soft-state black-hole binaries, showing a luminous soft component and a distinctly steep hard tail (e.g. Boller, Brandt & Fink 1996). For this reason, it is thought that NLS1s might be the analogue of the soft-state black-hole binaries at high accretion rate. (However, see below.)

2.2 Time Variabilities

A close similarity between AGN and black-hole binaries is found also in the characteristics of time variabilities. Both Seyfert 1s and the hard-state black-hole binaries exhibit rapid large-amplitude intensity fluctuations (flickering), though the origin of the flickering still remains unknown. The shape of the power density spectrum appears to be similar between AGN and black-hole binaries in the hard state, if the frequency is scaled. The power density spectrum is roughly a power law (fractal nature) down to a certain frequency and then turns flat at lower frequencies. The turn-over frequency is orders of magnitude lower for Seyferts than for black-hole binaries, which might reflect the difference in black hole mass. Attempts have been made to estimate the black hole masses of AGN by comparing the normalized power density spectra with that of a black hole binary whose compact object mass has been known (Hayashida et al. 1998), although the underlying physics is not understood.

NLS1s show a different type of fluctuations. NLS1s exhibit the most violent intensity variation among AGN with short time scales of hours or less. Not only the hard component, but sometimes the soft component also varies. On the contrary, the soft component of black-hole binaries in the soft state is fairly stable, and the hard tail, though it varies by a large factor, changes relatively slowly with time scales not much shorter than an hour. Obviously, a scaling law with the black hole mass does not hold here. Either the analogy of NLS1s with the soft state black-hole binaries is not valid, or the cause of the intensity variation is different between them.

2.3 Relativistic Jets

Table 1. Sources of Relativistic Jets in the Galaxy
(data taken from Mirabel & Rodríguez 1999)

Source	Compact object	V_{app} [2]	V_{int} [3]
GRS 1915+105 *	BH	$1.2c - 1.7c$	$0.92c - 0.98c$
GRO J1655-40 *	BH	$1.1c$	$0.92c$
XTE J1748-288 *	BH	$1.3c$	$> 0.9c$
SS 433	BH or NS	$0.26c$	$0.26c$
Cyg X-3 *	BH or NS	$\sim 0.3c$	$\sim 0.3c$
CI Cam *	NS ?	$\sim 0.15c$	$\sim 0.15c$
Sco X-1	NS	$\sim 0.5c$	
Cir X-1 *	NS	$\geq 0.1c$	$\geq 0.1c$
1E 1740.7-294	BH ?		
GRS 1758-258	BH ?		

* during transient outburst

Both AGN and Galactic X-ray binaries can produce relativistic jets. Despite a huge difference in the system size, it is probable that the jet production mechanism is fundamentally the same for both systems. The dynamical evolution of jets can be studied much more in detail for X-ray binaries, because of the short distances to the source and of far shorter time scales than in AGN. Superluminal jets have been observed from three transient black-hole binaries during the early phase of outbursts. In these cases, plasma blobs were ejected at a highly relativistic speed when a sudden accretion occurred and the luminosity increased near to the Eddington limit.

On the other hand, the jets of the radio-loud quasars seem to be sustained at much lower luminosities than the Eddington limit. If the interpretation of the spectral similarity is right, their power-law spectra similar to the hard-state black-hole binaries suggest low accretion rates. This feature apparently breaks down the parallels between AGN and black-hole binaries discussed in Section 1.1.

2.4 Reflection Component

If an optically-thick cool accretion disk is present and illuminated by X-rays, part of the incident X-rays will be reflected. The presence of a reflection component was predicted by Lightman & White (1988) and first detected by Pounds et al. (1990) from the X-ray spectra of Seyfert 1s obtained with *Ginga*. The reflection component is also confirmed in X-ray binaries, and currently studied intensively with respect to the accretion disk structure. The spectrum of the reflection component is characterized by an extremely hard spectrum (since photoabsorption dominates below 10 keV) with an intense fluorescent Fe-K line, and a broad bump (hard X-ray bump) around 20 keV. The reflection component accompanied by the Fe-K line from AGN is considered to be the evidence for the presence of an accretion disk.

3 Relativistically Broadened Iron Lines

Evidence for AGN being supermassive black holes has been obtained from radio and optical observations that measure the Kepler velocity around the central object, hence estimating the mass. The presence of a supermassive object at our Galactic center was also concluded from the measurement of proper motions of stars near the Galactic center (Eckart & Genzel 1996). While these results are convincing, they are probing motions at large distances of the order $\sim 10^5 r_g$, and require separate arguments for the compactness of the object.

If the large (special & general) relativistic effects that are unique to the direct vicinity of black holes were discovered, it would provide the most direct proof of a black hole. Since X-rays come from the inner part of the disk close to the central object, the X-ray spectrum is the best to look for such relativistic effects. The profile of the Fe K-line emitted near a black hole was calculated earlier by Fabian et al. (1989). The line is expected to be skewed (a blue horn and a weaker red tail), very broad and gravitationally redshifted.

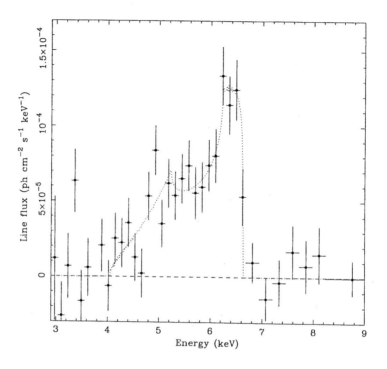

Fig. 3. Relativistically broadened Fe-K line in MCG 6-30-15. The dotted line shows the best-fit profile from the model of Fabian et al. (1989)

Such a relativistically broadened Fe K-line (the rest-frame energy of 6.4 keV) was first observed from the Seyfert 1 galaxy MCG 6-30-15 with *ASCA* (Tanaka et al. 1995), as shown in Fig. 3. The observed line profile is in good agreement with the expected one. From the subsequent observations of other Seyfert 1s, it is found that they commonly show similar broadened Fe-K lines (Nandra et al. 1997a), although a narrow Fe-K line (6.4-keV) emitted from distant parts of the disk may also be present. Broad Fe K-lines were also found in Seyfert 2s.

An interesting fact is found in MCG 6-30-15 that the line profile changes from time to time (Iwasawa et al. 1996). In an extreme case, the blue horn is strongly suppressed, and the profile shows the feature expected for an extreme Kerr hole. Iwasawa et al. (1999) suggest that X-rays illuminate the disk locally, and that the peculiar line profile is caused when the line emitting region is very close to a Kerr hole.

There is a clear tendency of decreasing Fe-K line intensity with increasing AGN luminosity (Nandra et al. 1997b). QSOs emit much weaker Fe-K line than Seyferts, independently of radio loudness. This is interpreted as due to the effect of ionization of the accretion disk when the luminosity is high.

4 Unified Model

The unified model of AGN was first introduced by Antonucci & Miller in 1985. The presence of a torus plays a dominant role on the observed properties of AGN, depending on the viewing angle with respect to the axis of the torus. Seyfert 2s had been suspected to be the case where the nucleus and the broad line region are obscured by such a torus. In agreement with this picture, X-ray observations of Seyfert 2s revealed the obscured nuclei with X-ray spectra showing heavily absorbed power-law component (Awaki et al. 1991). Because of the heavy absorption at low energies, the apparent X-ray luminosities of Seyfert 2s are much lower than Seyfert 1s, typically $\sim 10^{-2}$ of Seyfert 1s. In fact, if the absorption is corrected for, the intrinsic spectra are similar to those of Seyfert 1s and the luminosities are also of the same order of magnitude as Seyfert 1s. Another characteristic spectral feature of the Seyfert 2 spectra is an intense Fe-K emission line.

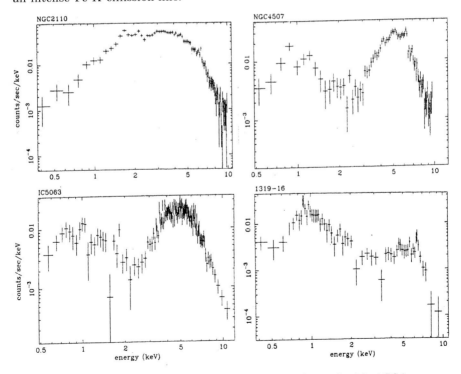

Fig. 4. Examples of the spectra of Seyfert 2s observed with *ASCA*

The recent *ASCA* observations allowed detailed study of Seyfert 2 X-ray spectra (e.g. Ueno 1997; Awaki et al. 2000). Some of the spectra are shown in Fig. 4. A remarkable feature is the common presence of a less-absorbed soft component, which becomes visible because of the low-energy attenuation of the power-law nuclear component by absorption, presumably through a molecular torus. The soft components are considered to be a mixture of the scattered nuclear component by the ionized gas surrounding the nucleus (consistent with the optical polarization), and an extended thermal emission from the host galaxy. The soft components often show complex spectra including the features of photoionization. Future high-resolution spectroscopic observations will tell us a great deal.

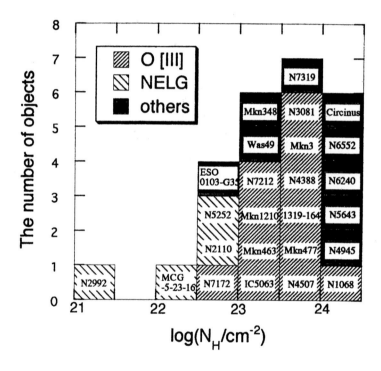

Fig. 5. Distribution of the observed absorption column density (Ueno 1997)

Seyfert 2s show a different amount of absorption from source to source, with the column density N_H ranging from 10^{22} to $> 10^{24}$ H-atoms cm^{-2}, as shown in Fig. 5 taken from Ueno (1997). With *ASCA*, for its limited energy band up to 10 keV, larger N_H values than 3×10^{24} cm^{-2} cannot be determined accurately, since in these cases a heavily absorbed nuclear component does not show up significantly below 10 keV. Such examples (NGC 1068 and NGC 6240) are shown in Fig. 6.

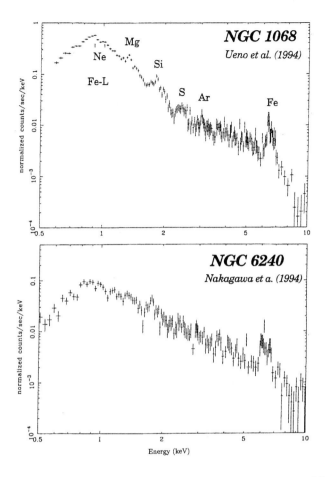

Fig. 6. Spectra of NGC 1068 and NGC 6240 observed with *ASCA*

The observed N_H distribution suggests the presence of cases of even heavier absorption. Hard X-ray observations above 10 keV are therefore very important. In fact, observations with *BeppoSAX* PDS detected the presence of luminous nuclear component from NGC 1068 (Matt et al. 1997) and NGC 6240 (Vignati et al. 1999), as shown in Fig. 7.

Whether or not heavily absorbed, so-called type 2 QSO exist as a class is still an issue. If they exist, they would be much fainter in X-rays than ordinary QSO. Good candidates to look for would be the infrared-luminous galaxies. In fact, the recent X-ray observations with *ASCA* and *BeppoSAX* found that several ultraluminous infrared galaxies (ULIRG) harbor hidden AGN of quasar-level X-ray luminosities. As listed in Table 2, at least three ULIRGs show intrinsic L_X in excess of 10^{44} erg s^{-1} in the 2–10 keV band. In those cases where the absorption is particularly heavy so that no signature

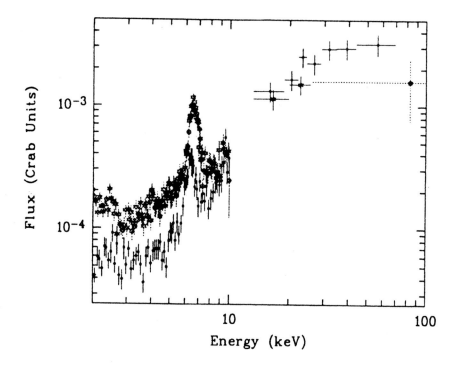

Fig. 7. Wide band spectra of NGC 1068 and NGC 6240 observed with *BeppoSAX*

Table 2. Infrared and X-ray luminosities of ULIRGs

Name	z	$\log(L_{FIR}/L_\odot)$	$\log(L_{X,\ 2-10keV}/L_\odot)^*$
NGC 1068	0.004	11.1	10.3
NGC 6240	0.024	11.5	11.0
Mrk 273	0.037	11.8	9.1
Mrk 231	0.042	12.1	8.9
IRAS 05189−2524	0.042	11.7	9.9
IRAS 08572+3915	0.058	12.1	8.3
IRAS 23060+0505	0.174	12.5	10.6
IRAS 20460+1925	0.181	11.8	10.5

* $L_{X,\ 2-10keV}$: X-ray luminosity in 2–10 keV corrected for absorption.
Data taken from Nakagawa et al. (1999) except NGC 1068 (Matt et al. 1997), NGC 6240 (Vignati et al. 1999) & IRAS 23060+0505 (Brandt et al. 1997)

of nuclear component is visible, the X-ray luminosity may still be underestimated. Hard X-ray observation above 10 keV is required for obtaining the correct luminosity. Evidently, the X-ray samples currently available are limited to nearby ULIRGs. Many more hidden QSOs may be found with the new X-ray observatories.

Incidentally, what powers ULIRGs is under debate. According to Genzel et al. (1997), majority of the ULIRGs are powered by starburst activities and a minor fraction of them by AGN activities. As seen in Table 2, the AGN contribution for some of them is large enough to account for the infrared luminosity, if one takes into account the bolometric X-ray luminosity and the UV luminosity of AGN. On the other hand, there are certainly cases for which the AGN power is far insufficient, $\log(L_{X,\ 2-10keV}/L_{FIR}) < 10^{-3}$, to account for the far-infrared luminosity. However, the present sample is still too small to draw a meaningful conclusion, and possibly subject to selection bias. It is likely that more cases of heavily-absorbed powerful AGN, such as NGC 6240, will be found in the near future. In addition, it may well be possible that the starburst activity and the nuclear activity are related to each other.

5 Cosmic X-Ray Background

Setti & Woltjer pointed out the importance of the AGN contribution to the CXRB already in 1973. They also presented in 1989 that a significant contribution of heavily absorbed AGN could well solve the so-called "spectral paradox", i.e. the CXRB spectrum is much harder than those of non-absorbed AGN.

Based on this idea, Comastri et al. (1995) demonstrated that the observed CXRB spectrum can be reproduced by an integrated contribution of AGN, considering a wide distribution of absorbing column N_H. In fact, the *ASCA* spectral observations show an increasing population of Seyfert 2s and possibly type 2 QSO, which show hard spectra as we go to lower flux levels. These hard-spectrum AGN will most probably resolve the "spectral paradox". Thus, the current scenario that AGN account most, if not the whole, of the CXRB seems to be well in order.

Yet, an unresolved problem still remains. This is the long-known high-energy break around 20–40 keV. The shape of the break has been accurately measured (Marshall et al. 1980), which is sharp as described by an exponential cut-off. The right model should precisely reproduce the observed shape of the break. This gives a fairly stringent constraint on the spectra and the redshift distribution of the constituents that are the major contributor near the break energy. Seyfert galaxies are known to show a high-energy cut-off, but not all at the right energy. The available samples of QSO spectra are limited, yet those of luminous QSOs do not show a high-energy cut-off. Extensive studies of AGN above 10 keV are still required for the full understanding of the properties of AGN and the CXRB.

References

1. Antonucci, R.R.J., Miller, J.S. (1985) Astrophys. J. 297, 621
2. Awaki, H., Koyama, K. et al. (1991) Publ. Astron. Soc. Japan 43, 195
3. Awaki, H., Ueno, S. et al. (2000) astro-ph/0006054
4. Boller, T., Brandt, W.N., Fink, H. (1996) Astron. Astrophys. 305, 53
5. Brandt, W.N., Fabian, A.C. et al. (1997) Mon. Not. R. Astron. Soc. 290, 617
6. Comastri, A., Setti, G. et al. (1995) Astron. Astrophys. 296, 1
7. Eckart, A., Genzel, R. (1996) Nature 383, 415
8. Fabian, A.C., Rees, M.J. et al. (1989) Mon. Not. R. Astron. Soc. 238, 729
9. Genzel, R., Lutz, D. et al. (1998) Astrophys. J. 498, 579
10. Hayashida, K., Miyamoto, S. et al. (1998) Astrophys. J. 500, 642
11. Iwasawa, K. Fabian, A.C. et al. (1996) Mon. Not. R. Astron. Soc. 282, 1038
12. Iwasawa, K. Fabian, A.C. et al. (1999) Mon. Not. R. Astron. Soc. 306, L19
13. Lightman, A.P., White, T.R. (1988) Astrophys. J. 335, 57
14. Marshall, F.E., Boldt, E.A. et al. (1980) Astrophys. J. 235, 4
15. Matt, G., Guainazzi, M. et al. (1997) Astron. Astrophys. 325, L13
16. Mirabel, I.F., Rodríguez, L.F. (1999) Ann. Rev. Astron. Astrophys. 37, 409
17. Nakagawa, T., Kii, T. et al. (1999) Astrophys. Space Sci. 266, 43
18. Nandra, K., George, I.M. et al. (1997a) Astrophys. J. 477, 602
19. Nandra, K., George, I.M. et al. (1997b) Astrophys. J. 488, L91
20. Pounds, K.A., Nandra, K. et al. (1990) Nature 344, 132
21. Setti, G., Woltjer, L. (1973) in Proc. IAU Symp. No. 55, 1972, "X- and Gamma-Ray Astronomy", eds. H. Bradt and R. Giacconi, Reidel, Dordrecht, p.208
22. Setti, G., Woltjer, L. (1989) Astron. Astrophys. 224, L21
23. Sunyaev, R.A., Titarchuk, L.G. (1980) Astron. Astrophys. 86, 121
24. Tanaka, Y., Nandra, K., Fabian, A.C. et al. (1995) Nature 375, 22
25. Tanaka, Y., Shibazaki, N. (1996) Ann. Rev. Astron. Astrophys. 34, 607
26. Ueno, S. (1997) ISAS Research Note RN619
27. Vignati, P., Molendi, S. et al. (1999) Astron. Astrophys. 349, L57

The Role of Ground-Based Infrared Astronomy

Alan Moorwood

European Southern Observatory
Karl-Schwarzschild-Str. 2
D-85748 Garching

Abstract. Infrared observations from the ground can contribute to our understanding of the AGN and related phenomena in various, sometimes unique, ways. The aim of this brief review is to illustrate this using results obtained at various ESO telescopes, with emphasis on those now coming from ISAAC at the VLT which are of interest both in their own right and as advertisements for the future prospects opened up by this new class of 8m telescope.

1 Introduction

Although AGN are powered intrinsically by high energy phenomena, much of their luminosity emerges in the infrared following re-processing by dust. In the 'standard' model this includes 'hot' dust located in the torus within a few parsec of the nucleus and colder dust further out in the ISM of the host galaxy. The infrared wavelength range is also rich in spectral features which provide unique diagnostic information. Examples in just the near infrared covered briefly in this review include 'coronal' lines from highly ionized species which can be used to both identify obscured AGN and probe their EUV continuum shapes; hydrogen recombination and molecular hydrogen lines excited by both the AGN and surrounding star formation regions and stellar absorption features such as CO and SiI which provide information on the stellar content. Advances in infrared instrumentation plus the fact that many AGN's are heavily obscured in the visible is also increasingly favouring the use of infrared rather than visible spectroscopy for more traditional studies e.g of the gas kinematics. The advent of the new generation of 8-10m class ground-based telescopes has also now revolutionized our ability to follow the visible rest frame spectra of AGN and star forming galaxies out to z = 3 and more. Some illustrative examples are briefly discussed in this review with special emphasis on results obtained at ESO, particularly those obtained recently with ISAAC at the VLT. The overview given here is personal and far from complete. It is worth noting for example that near infrared surveys are actually now a source of new AGNs. Looking ahead, perhaps the most exciting prospect is that the interferometric capabilities planned for the VLT and other large telescopes will provide unprecedented spatial resolution down to the masec range in the not too distant future.

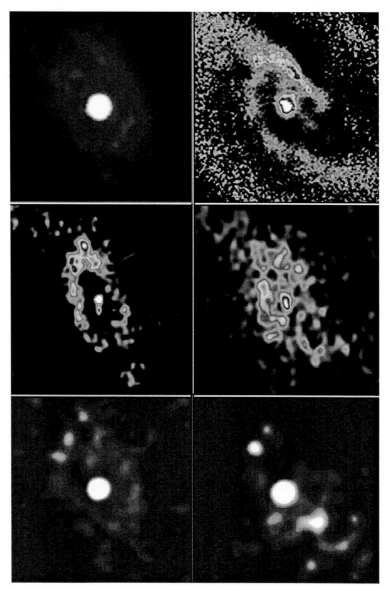

Fig. 1. Infrared images of the central 25x25″ of NGC1365. N is at the top and E to the left. Panels are: top (IRAC/2.2m) Left K′(2.16μm) Right J-K′: middle (IRAC + Fabry Perot/2.2m) Left Brγ Right H$_2$: bottom Left (IRAC/2.2m) L(3.8μm) Right (TIMMI/3.6m) N(10μm).

2 Seyfert Galaxies

2.1 An Infrared View

Figure 1 shows a mosaic of images of the Seyfert 1 galaxy NGC1365 in several continuum bands and emission lines of specific interest obtained with the IRAC and TIMMI cameras on the ESO 2.2m and 3.6m telescopes respectively. In contrast to K' it is interesting how clearly the inner spiral structure is visible in the J-K' image and that it appears to connect with and thus provide a possible mechanism for fuelling the AGN. The Brγ hydrogen recombination line emission shows a clear ring of star formation about $\sim$ 1kpc across and provides quantitative information on the UV continua and star formation rates in the individual star forming clusters. This probably corresponds to the position of the inner Lindblad resonance and may be unrelated to the AGN activity. The molecular hydrogen emission is somewhat more complicated because it is partially excited by star formation driven shocks and partially by either shocks or x-ray emission from the AGN. At the longer wavelengths, in the thermal infrared, the 3.8μm and 10μm images show more deeply embedded star clusters as well as emission at the position of the AGN. Within the 'standard' model this is mostly emitted by dust heated to a few × 100K within a surrounding $\sim$ pc size torus. Even diffraction limited images at 4m class do not spatially resolve this emission and only set an upper limit of typically $\sim$10pc for the size of the emitting region. Interferometry at these wavelengths with VLTI however should be capable of resolving it and thus establish whether or not it is in the form of a torus in a few nearby Seyferts.

2.2 Red Supergiants as Star Formation Tracers

Star formation is often present in the central regions of Seyfert galaxies although it remains statistically unclear whether or not its incidence is higher than in the spiral galaxy population as a whole. Its relationship, if any, with the AGN activity also remains unclear. Star formation close to or at the nucleus may also not be easily revealed by gas emission line tracers such as those discussed above. Red supergiants therefore offer an interesting alternative because they also trace young stellar populations and can be detected spectroscopically in the infrared H and K bands where they exhibit strong CO, SiI and other characteristic absorption features. The main difficulties are the possible confusion with red giants which show the same, but usually weaker, features and dilution of the features by non-stellar continuum in the case of Seyfert galaxies. With adequate spectral resolution, however, these problems can be circumvented by using the same infrared spectra to derive the stellar M/L ratio which is extremely sensitive to the red supergiant content. Such observations have been made of samples of both Seyfert and normal galaxies by Oliva et al.(1999) using IRSPEC at the ESO NTT to test for the presence of star formation close to the nucleus in the Seyferts

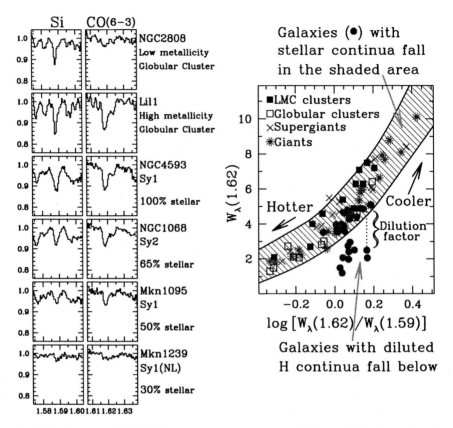

Fig. 2. Left panel: IRSPEC spectra around 1.6μm of SiI and CO absorption in globular clusters and Seyfert galaxies. Right panel: Behaviour of $W_\lambda(CO)$ versus $W_\lambda(CO)/W\lambda(SiI)$ for giants, supergiants, globular clusters and galaxies showing the effect of dilution by the non-stellar continuum in Seyfert galaxies.

and for any differences between the types 1 and 2. In a continuation of this programme, higher spatial resolution spectra have recently been obtained of several Seyferts with ISAAC at the VLT but the analysis is still in progress. The technique is illustrated in Fig. 2. The left panel shows IRSPEC spectra around 1.6μm of the SiI and one of the CO features in globular clusters and Seyfert galaxies. The right hand panel shows the well behaved relationship observed between CO(1.62μm) EW and CO/SiI EW ratio for giants, supergiants, globular clusters and galaxies. Well behaved that is except for Seyfert galaxies whose CO EW is diluted by the non-stellar continuum. Because the CO and SiI features are close in wavelength their ratio is not affected by dilution. The latter can therefore be estimated for individual galaxies from this plot and used to determine the true stellar 1.6μm continuum flux and luminosity. The total mass within the same region can then be determined

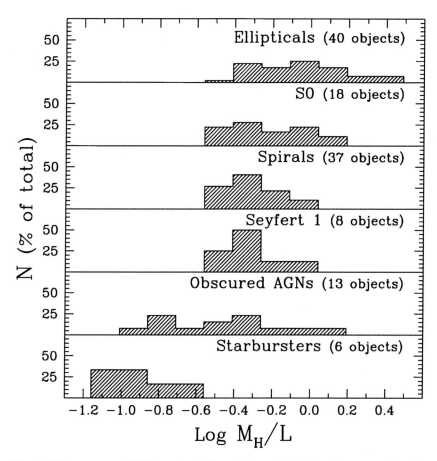

Fig. 3. Histograms of M/L ratios deduced from the infrared spectra of various galaxy types. The ratio is about an order of magnitude lower in starburst than in elliptical galaxies due to the presence of red supergiants. Whereas the distribution in Seyfert 1s is indistinguishable from normal spiral galaxies the M/L distribution in Seyfert 2s is wider and extends down to almost the lowest values observed in starburst galaxies.

from the measured velocity dispersion. The result to date is shown in the form of histograms of M_H/L in Fig. 3. The extremes in M/L ratio are exhibited by ellipticals shown at the top and starburst galaxies shown at the bottom whose values are $\simeq$ a factor 10 lower. Values in the target Seyfert sample lie in between. Whereas the Seyfert 1 distribution is indistinguishable from that of normal spirals, however, that of 'obscured AGNs' (Seyfert 2) is much wider and extends down to the range occupied by the starbursters. This is not expected within the standard Seyfert unification scheme in which the difference in observed properties is determined solely by the inclination of the torus to the line of sight. It could be explained by the fact that AGNs in starburst galaxies are more likely to be classified as Seyfert 2's because of extinction by the star forming clouds themselves. It could also, however, be taken as support for the evolutionary scenario in which gas flow towards the nucleus first triggers star formation which fuels the black hole resulting in the appearance of Seyfert 2 and finally Seyfert 1 characteristics as the starburst fades and the line of sight to the AGN clears.

2.3 Probing AGN EUV Continua with IR Coronal Lines

The infrared spectral region is relatively rich in ionic lines from species with ionization potentials $\geq$ 100eV. The discovery spectrum containing such a line - [SiVI](1.96μm;166eV) - was of the Seyfert 2 galaxy NGC1068 obtained with the IRSPEC spectrometer mounted on the ESO 3.6m telescope (Oliva & Moorwood, 1990). This was followed by [SiVII](2.49μm; 205eV) detected in the same galaxy and [SiVI], [SiVII], [SiIX] (3.94μm; 303eV), [SIX](1.25μm;328eV) and [CaVIII](2.325μm;128eV) in the Circinus type 2 Seyfert galaxy with IRSPEC at the ESO NTT telescope (Oliva et al. 1994). This new and richer coronal line spectrum prompted first a return to the question of their origin which was initially addressed, but not satisfactorily resolved, in relation to the [Fe] coronal lines in the visible. The best fit models tend to favour photoionization of moderate density gas by a hard EUV spectrum rather than e.g shock heated gas (Oliva et al. 1994, Marconi et al. 1996). It is worth noting that the fact that the coronal line widths are only $\simeq$ 100km/s broad in the Circinus galaxy spectrum essentially excludes heating by fast shocks without recourse to models. Inverting photoionization models in the case of the Circinus galaxy also yields an essentially flat EUV continuum spectrum (Oliva et al. 1994) whose extrapolation is consistent with the x-ray spectrum subsequently measured with the ASCA satellite (Matt et al. 1996). Although strictly outside the scope of this review it should be noted for completeness that several lower excitation lines at longer wavelengths which probe the softer UV continuum have also been observed in Seyfert galaxies by the ISO satellite. At least in the case of the Circinus galaxy, photoionization modelling of these requires the presence of an additional strong UV 'bump' around 70eV which is consistent with the intrinsic continuum shape of Seyfert 1's and quasars deduced by interpolating direct UV and X-ray observations (Moorwood et al.1996).

3 Weighing Black Holes with the VLT

Infrared spectroscopy offers a powerful technique for tracing gas motions and hence of estimating the black hole masses in AGN whose central regions are often heavily obscured in the visible. A prime example is Centaurus A on which this approach has been recently demonstrated by Marconi et al.

(2000) who used the ISAAC spectrometer on the ESO VLT (Moorwood et al. 1999) to map the inner velocity field in the lines of Paβ, [FeII], Brγ and H$_2$ at R$\simeq$10000 and with $\simeq$ 0.5″ spatial resolution. They find three distinct kinematical systems i) a rotating 'nuclear disk' of ionized gas confined to the inner 2″ ii) a ring like system of molecular gas with a $\simeq$ 6″ inner radius detected only in H$_2$ and iii) a normal extended component of gas rotating in the galactic potential. The nuclear disk is in keplerian rotation around a central mass concentration which is dark (M/L $\geq$ 20 M$\odot$/L$_K\odot$), point-like (R $\leq$ 4pc) and thus most probably a supermassive black hole with a dynamical mass of $\simeq$ 2x10^{8}M$\odot$.

4 High z Radio Galaxies

The availability of high sensitivity infrared spectrometers on large ground-based telescopes has now opened up the possibility of following the rest frame visible studies of radio and other galaxies out to higher redshifts. Fig 4 shows a 1-2.5μm spectrum of the z=2.4 radio galaxy MRC0406-244 obtained during the commissioning of ISAAC on the ESO VLT. At this redshift, the infrared spectrum corresponds to the rest frame visible and contains the familiar, prominent, emission lines [OII], Hβ. [OIII], Hα, [OI] and [SII]. Also of interest is the detection in this spectrum of continuum emission around 2μm consistent with the presence of a stellar continuum. In a more extensive study of a sample of 6 radio galaxies Fosbury (2000) has found strong continua and extremely broad ($\simeq$ 10000km/s) Hα lines in 3. Detailed studies of the z=2.4 radio galaxy MRC1138-262 over the complete rest frame wavelength range from 1000-8000Å (combining ISAAC/VLT and Keck spectra) shows that its spectrum can be well modelled by combining an elliptical galaxy spectrum with direct and scattered contributions from a quasar which is obscured along the line of sight. As already suspected, therefore, at least some radio galaxies do appear to be 'edge on' quasars in elliptical galaxy hosts. Also of interest in the case of the MRC0406 observation is that three other galaxies were observed simultaneously with the radio galaxy. Two of these are members of a small group of ERO's (extremely red objects) within a few arcsecs of the radio galaxy and suspected originally to be members of a cluster centred on the radio galaxy (McCarthy et al. 1992). Their spectra visible in the upper panels, however, turn out to be smooth with no spectral lines or breaks corresponding to the radio galaxy redshift. ISAAC spectra of a larger sample of EROs found in surveys with SOFI at the NTT and ISAAC at the VLT by Cimatti et al. (1999) have also failed to show emission lines or breaks and spectrophotometric fits suggest that they are elliptical galaxies at z $\simeq$ 1. Nevertheless, these authors confirm an overdensity of such objects in the direction of higher z radio galaxies which is difficult to understand except possibly as a selection effect caused by gravitational lensing.

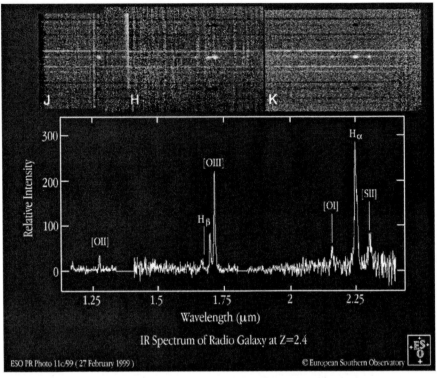

Fig. 4. ISAAC 1-2.5μm spectrum of the z=2.4 radio galaxy MRC0406-24. The wavelength range corresponds to the rest frame visible as can be seen from the familiar emission lines [OII], Hβ, [OIII], [OI], Hα and [SII]. The upper panels containing the 2D spectra show the spectra of three other galaxies observed simultaneously with the radio galaxy. (Each positive spectrum is accompanied by 2 negative ones as a result of the sky subtraction technique used). Two of these are EROs (Extremely Red Objects) whose spectra, however, do not show features or breaks and are probably at lower redshifts than the radio galaxy.

5 High z Star Forming Galaxies

Infrared spectroscopy with large ground-based telescopes has also revolutionized the study of star forming galaxies at high redshift which can now be observed in the same emission line tracers - [OII], Hβ, [OIII], Hα used to study nearby galaxies. ISAAC at the VLT has already been used to obtain spectra of samples of galaxies at $z = 0.5$-3.3. Hα is a particularly important star formation tracer which more directly measures the instantaneous star formation rate and suffers lower extinction than the UV continuum which has been adopted as the primary indicator of star formation rate versus redshift to date. Out to $z \simeq 2.5$ it is now possible to independently measure this using the Hα line - and to higher redshifts using Hβ. The range $z \simeq 2$-3 is

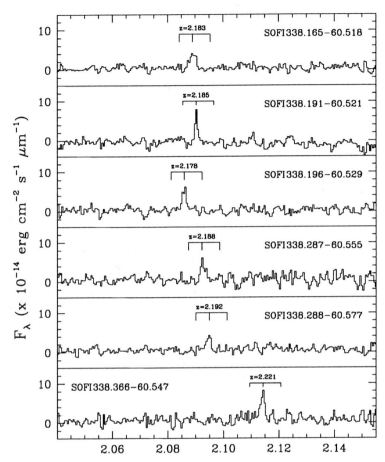

Fig. 5. ISAAC K band spectra of 6 candidate Hα emitting galaxies at z=2.2 found in a narrow band infrared survey. The expected emission line is clearly present and provides an estimate both of the star formation rate and mass of these galaxies.

also of specific interest because it corresponds to the peak of the space distribution of AGN - and potentially also the cosmic star formation rate density based on interpolations from lower and higher redshifts. It is a difficult range for optical studies, however, because of the difficulty of determining both spectroscopic and photometric redshifts. Our knowledge of the star formation rate density from the rest frame UV continuum has thus been largely confined to the ranges $z \leq 1$ and ≥ 3. One of the VLT ISAAC programmes has therefore specifically targeted Hα emitting galaxies at $z = 2.2$ (Moorwood et al. 2000). The sample was established in a pre-cursor narrow band 2 μm infrared imaging survey for z=2.2 Hα emission conducted using the SOFI instrument at the ESO NTT telescope. Out of about 10 candidate Hα emitting objects detected over an area of 100 sq. arcmin (including the HDFS),

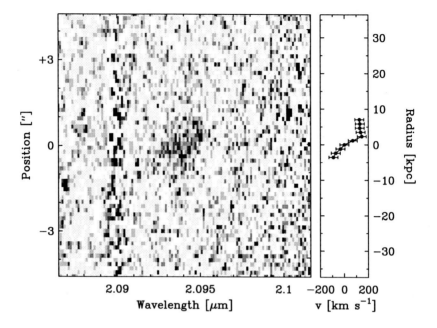

Fig. 6. The left panel is the 2D spectrum of one of the z=2.2 galaxies showing tilted Hα emission indicative of rotation. The right panel shows the extracted rotation curve whose velocity spread falls within the range observed in nearby spirals.

at least the 6 whose ISAAC spectra are shown in Fig. 5 do show an emission line at the survey wavelength which is most plausibly identified with Hα. It should be noted that these spectroscopic confirmations required integration times of only 1 hr even in relatively poor seeing conditions. Star formation rates are around 25M☉/yr. The spectra are at high enough resolution to also resolve the lines in most cases yielding velocity dispersions of 60-200km/s - indicative of relatively massive systems assuming the linewidths are dominated by gravity. Evidence that this is the case is provided by the spectrum of one object shown in Fig. 6 whose Hα emission is clearly tilted. Although the s/n ratio is low (rotation curves were neither *a priori* being searched for or expected to be seen !) the extracted rotation curve shown on the right implies a terminal velocity ≥140km/s which is within the range measured for nearby spiral galaxies. Although small, this is the largest sample of spectroscopically confirmed star forming galaxies at this redshift detected by narrow band infrared imaging. In relation to the above discussion therefore it is of great interest that the derived Hα luminosity function at z=2.2 is identical to that found at z ≃1.3 by Yan et al. (1999) from an Hα grism survey conducted with NICMOS on HST. The comparison is shown in Fig. 7 where the solid squares correspond to z=2.2 and the upper curve is a Schechter function fit

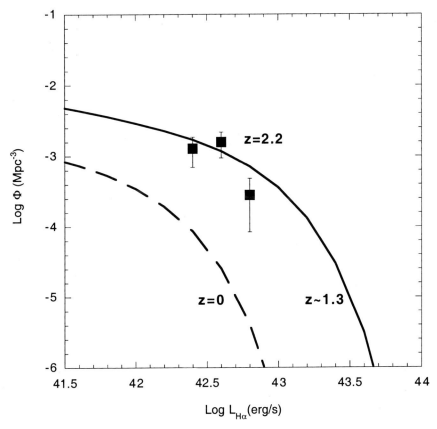

Fig. 7. Hα luminosity functions. The filled squares correspond to the z=2.2 SOFI/ISAAC sample. The solid curves are Schechter function fits to the data of Yan et al. (1999) at z ∼ 1.3 (solid) and Gallego et al. (1995) at z=0 (dashed).

to the z=1.3 data. For comparison, the lower curve is for z=0 (Gallego et al. 1995). Based on Hα alone therefore the strong evolution from z = 0-1 is confirmed but does not appear to continue to higher redshifts. The actual star formation rate density determined at z = 2.2 is also equal to the latest values at z ≃ 3 deduced from extinction corrected UV continuum measurements (Steidel et al. 1999) and which also remain roughly constant to z ≥ 4. In combination, therefore, there is no compelling evidence at the moment for a peak in the cosmic star formation rate density at the epoch of the AGN peak.

In another ISAAC VLT study of high redshift star forming galaxies Pettini et al. (2000) have obtained spectra of about 8 z ≃ 3.2 Lyman Break galaxies covering the [OII](3727Å), [OIII](4959,5007Å) and Hβ lines. All but one has been detected in [OIII] plus an additional galaxy was serendipitously discovered within a few arcsec and at almost the same redshift as one of the target sample. The one not detected in the Lyman Break survey remains a puzzle in that it exhibits almost the same [OIII] line flux but is several mags. fainter in the continuum. The reason for targeting the oxygen lines was to obtain a first idea of the metal abundances with the result that O/H $\simeq$ 0.1-0.3 in this sample. Again the lines also show velocity dispersions $\simeq$80 km/s which, computing M_B from the infrared continuum magnitudes, places them typically 3 mags. above the local Tully-Fisher relationship.

As an overall summary of the high z observations so far perhaps the most interesting general conclusion is that relatively well developed systems were clearly already in place at z $\simeq$ 2-3.

6 Acknowledgements

I wish to thank my various collaborators involved in the work presented here, particularly Tino Oliva, Paul van der Werf, Jean Gabriel Cuby and Max Pettini. I also wish to pay special tribute to Lo Woltjer without whose vision, encouragement and support this work would not have been possible because most of the instruments and telescopes featured here would probably never have been built.

References

1. Cimatti, A., Daddi, E., di Serego Aligheri, S., et al. 1999, A&A, 352, L45
2. Fosbury, R.A.E., 2000, SPIE 4005, 75
3. Gallego, J., Zamorano, J., Aragon-Salamanca, A., Rego, M. 1995, ApJ, 455, L1
4. Madau, P., Fergusen, H.C., Dickinson, M., et al. 1996, MNRAS, 283, 1388
5. Marconi, A., van der Werf, P.P., Moorwood, A.F.M, Oliva, E. 1996, A&A, 315, 335
6. Marconi, A., Capetti, A., Axon, D., et al. 2000 (AJ submitted)
7. Matt, G.,Fiore, F., Perola, G. C., et al. 1996, MNRAS, 281, L69
8. McCarthy, P.J., Persson, S.E., West, S.C. 1992, ApJ, 386, 52
9. Moorwood, A.F.M., Lutz, D., Oliva, E., et al. 1996, A&A, 315, L109
10. Moorwood, A.F.M. et al. 1999, The Messenger, 95,1
11. Moorwood, A.F.M., van der Werf, P.P., Cuby, J.G., Oliva, E. 2000 (SPIE 4005,55 and A&A submitted)
12. Oliva, E., Moorwood, A.F.M. 1990, ApJL, 348, L5
13. Oliva, E., Salvati, M., Moorwood, A.F.M., Marconi, A. 1994, A&A, 288,457
14. Oliva, E., Origlia, L., Maiolino, R., Moorwood, A.F.M. 1999 A&A, 350, 9
15. Pettini, M., Moorwood, A.F.M., Steidel, C.C., et al. 2000 (in preparation)
16. Steidel, C.C., Adelberger, K.L., Giavalisco, M., et al. 1999, ApJ, 519, 1
17. Yan, L., McCarthy, P.J., Freudling, W., et al. 1999, Ap. J., 519,L 47

What May We Learn from Multi-Wavelength Observations of Active Galactic Nuclei

Thierry J.-L. Courvoisier

INTEGRAL Science Data Centre, 16 ch. d'Ecogia,
CH-1290 VERSOIX, Switzerland
and
Observatoire de Genève, CH-1290 SAUVERNY, Switzerland

Abstract. We discuss how several of the questions that remain unclear on the physics of Active Galactic Nuclei may find elements of answers when using in the coming years the extraordinary set of instruments that will be flying simultaneously to observe in all bands of the electromagnetic spectrum. The choice of questions mentioned here is personal and not exhaustive.

1 What Do We Know About Active Galactic Nuclei?

Wondering what may be learned about Active Galactic Nuclei (AGN) in the coming years requires that one thinks about what is currently known and about the type of knowledge that is relevant. Indeed, AGN seem to be complex systems that need not be described at a level at which the general physical processes are hidden behind local and/or statistical fluctuations. The learning process may in some sense be similar to understanding the physics of the terrestrial atmosphere by doing meteorological observations. The aim in this case is not to describe the local weather but to extract from the complex descriptions of the observations the guiding principles. Clearly, what knowledge or understanding one wishes to obtain on any given subject is a matter of personal taste. Indeed many people, even among astronomers, might have only a limited interest in AGN. The following account is therefore a personal approach to the subject.

Among the facts that are well established now and documented in a number of books and reviews in recent years one finds:

The source of energy is the gravitational energy of matter falling into a massive black hole.
With this general principle one deduces two characteristics of AGN, namely an estimate of the mass of the central black hole through the Eddington luminosity (more appropriately, this estimate gives a lower boundary to these masses). This mass is found to be up to 10^{10} solar

masses. Also from the luminosity of the objects and from reasonable assumptions on the efficiency with which accreted matter can be made available to be radiated one deduces the rate at which matter must be accreted into the central black hole. Rates of up to 100 solar masses per year for the brightest nuclei are found.

The role of the rotation of the central black hole in the global energetics still needs be understood.

The emission processes through which energy is radiated are now mostly identified. They include synchrotron emission, emission of dust heated by ultraviolet and X-ray radiation fields, Compton processes (be they thermal or due to relativistic electron distributions) and emission from optically thick media (black body radiation). It is possible from the identification of the emission processes and the shape of the spectral energy distributions to deduce (some of) the physical conditions of the emitting material.

The state of the gas is known for the gas that emits the lines characteristics of most AGN types. Temperature, density and filling factors as well as the origin of the ionisation can be read from the spectra and provide a picture in which photoionisation plays a dominant role.

Aberrations and geometry play an important role through Doppler boosting of components related to relativistic jets and through absorption of some components that happen to lie behind dense clouds of dust and gas. These effects considerably modify the appearance of AGN and probably lead to our AGN classification scheme.

AGN sit in the center of galaxies with which they must have intense relations, be it only through material that is falling into the galactic potential well to the AGN.

The evolution of AGN is well documented. The AGN number density and/or luminosity have greatly changed over the age of the Universe.

AGN have a very rich phenomenology for which we have large amounts of data. We thus have descriptions of many spectral energy distributions, of many types of variations in different objects and at different wavelengths and we have a vast zoology of objects in which populations differ greatly in one or several characteristic parameters. We are not always successful in understanding this rich phenomenology in terms of underlying physical processes.

2 Current Questions

2.1 Reprocessing

We know that reprocessing of radiation by a medium takes place in AGN. Examples are the emission of heated dust (reprocessed UV radiation), Compton processes in which soft photons acquire additional energy from hot electrons and the presence of a fluorescence Fe line among others.

What still needs be quantitatively assessed is the relative importance of the so-called Compton reflection hump that was suggested by Pounds et al. [20] to be important between 10 and some tens of KeV. This hump is due to the reflection of a primary X-ray component (taken as a power law) by cold material. That this causes a hump can be qualitatively understood by noting that the primary component is absorbed by the medium at low energies (the medium is assumed optically thick) while at high energies the scattering cross section decreases due to Klein-Nishina effects; see Mushotzky, Done and Pounds [15] for a review. There will therefore be a region in the spectrum where the reflection is maximal, thus producing a broad hump on top of the primary component.

It is suggested in models of the ultraviolet emission of AGN (see next section) that the UV emission is due to a disk heated by the primary X-ray component rather than by gravitational energy released within the (optically thick) accretion disk. In this scenario the reflection hump observed in the medium energy X-rays, the fluorescence line and the UV emission are all due to the same reprocessing scheme. All three elements should be consistently observed and variations in these components must be closely correlated and related (albeit possibly in a complex way) to the variations of the primary X-ray component.

Observations spanning the ultraviolet to hard X-ray domain can provide measurements that will allow us to separate the primary component from the reprocessed ones, to measure the flux in all components as a function of time and to seek some geometrical information by looking at the relative intensity of the components in different classes and as a function of time.

These observations will be possible using XMM and INTEGRAL in a coordinated manner over the coming decade on a variety of objects.

2.2 The Blue Bump

The peak in emission in the blue-ultraviolet region of the spectrum that is observed in Seyfert galaxies and quasars has been associated with the accretion process since 1978 [23]. This suggestion has given rise to a large activity in the subsequent years to provide a good match between observations and accretion disk theory. Despite 2 decades of work the concordance between the observations and theory is still not satisfactory. The continuum spectral energy distribution, spectral features, polarisation, relationship between continuum shape and luminosity and, possibly most importantly, the variability properties of the blue bump do not match those expected from an optically thick medium inside which gravitational energy is released to be radiated at the surface of the medium; see Courvoisier and Clavel [5] and the review on AGN accretion disk related problems in Koratkar and Blaes [14].

The blue bump flux is observed to vary almost simultaneously at all wavelengths from the visible to the ultraviolet range in 3C 273 as well as in Seyfert galaxies [5]. This is in contradiction with the expectations that disturbances that propagate through the accretion disk (or whatever other structure) do so on viscous timescales that are many order of magnitude longer than the observed lags or upper limits thereof. This has lead to the development of models in which an X-ray source shines onto the disk. The latter is thus heated from the outside and the illuminated regions respond to variations of the primary source on timescales that are given by light propagation effects in good agreement with measurements of lags [17]. One of the most elaborate models of this type is that of Haardt, Maraschi and Ghisellini [13] in which a multiple corona shines onto the disk. It may be worth mentioning that the observations require only that the signal responsible for the perturbances propagate with the speed of light, not necessarily that the energy be distributed with the speed of light.

In a detailed study of the visible and ultraviolet emission of 3C 273 Paltani Courvoisier and Walter [17] have shown that the short wavelength emission of the blue bump in 3C 273 can be well modeled by a series of independent events occuring at a rate of about 3 per year and having each an energy of some 10^{52} ergs. If this approach is physically sound (a good mathematical description of the data is not sufficient to ascertain the physical validity of the assumptions) than several questions must be answered: What is the origin of the events (see next subsection for a proposal)? What about the long wavelength emission of the blue bump? Can the analysis be generalised to other objects?

Be it in the frame of the accretion disk models with or without a corona or in that of the individual events, there are still fundamental questions to be solved to explain the nature of the blue bump emission.

2.3 How Does the Accretion Proceed?

Accretion disks [22] are a natural gas accretion structure in that they provide a means of expelling angular momentum while the material is being accreted provided that the nature of the viscosity is understood.

The difficulties of the accretion disk models to explain the observed properties of the blue bump, the description of the blue bump in 3C 273 in terms of individual events and the presence of stars in the center of galaxies have led us to study a model in which the accreted material is not a diffuse gas but rather in the form of stars [6], [24]. Stars orbiting a supermassive black hole at some 100 gravitational radii have velocities close to 0.1 c. When two such stars collide their kinetic energy amounts to some 10^{52} ergs for stars of the mass of the Sun. This is in good agreement with the energy contained in one event as discussed in the previous section.

In order to have few collisions per year (to provide the event rate described above) and to explain the average luminosity of the quasars there must be approximately the same mass in the central black hole and in the stellar population within some 100's of gravitational radii. In this volume the star density must be therefore very high, of the order of $10^{11} - 10^{14}$ stars per cubic parsec.

Accretion of matter in the form of stars and dissipation of the kinetic energy through stellar collisions rather than in a disk has the property that the dependence of the variability on the object luminosity discussed in Paltani and Courvoisier [18] based on IUE observations can be explained. This dependence is much flatter than expected based on independent events that are Poissonian distributed in time, suggesting that the event properties are a function of the average luminosity of the object. The star accretion scenario provides this naturally, because the average star collision energies depend on the distribution of stars with distance to the black hole, which also determines the average luminosity of the object [24].

Much work is still needed to assess whether this accretion process has a good prospect to contribute to the luminosity of AGN in a significant way. The emission processes occuring during and after a collision must be calculated and compared with observations of spectral energy densities and variability, the dynamics and evolution of very dense clusters in the vicinity of a massive black hole must be studied. An alternative line of work may be to prove that this model is irrelevant by showing that the accretion material is organised in a single plane and rotating all in the same direction.

2.4 Links Between Starburst and AGN Activities?

Starburst activity is observed in relation with AGN (the luminosity of which comes from accretion processes). This type of link can intuitively be expected. As the material falls in the central regions of a galaxy it will dissipate angular momentum. This will stir the interstellar material and provide conditions that are propitious for star forming activity.

Starburst activity with a luminosity that is comparable to that of the AGN seems to be common among absorbed Seyfert galaxies (Seyfert II galaxies). Both Oliva et al. [16] and Delgado and Heckman [7] report that at least close to half of the Seyfert II galaxies that they observed have important starburst activity. On the other hand, Oliva et al. [16] report that none of the 8 unobscured Seyfert galaxies (Seyfert I galaxies) they have observed showed starburst properties.

That the difference between Seyfert I and Seyfert II galaxies lies in the presence of absorbing material that obscures the nuclei of the Seyfert II galaxies is in no doubt based on the observation of broad lines in the polarised light of Seyfert II galaxies [4] and the clearly absorbed X-ray emission of the same.

The observations described above, suggest further that the presence of the obscuring material is related to the star forming activity. This link may indeed be expected as star formation requires the presence of large quantities of gas and dust forming dense clouds that are also associated with higher probabilities of obscuration.

The observations discussed here are in marked contrast with the expectations of the unified models [3] which postulate that Seyfert I and Seyfert II galaxies are differentiated by their orientation with respect to the line of sight. The model suggested by the different level of starburst activity in both types of galaxies is indeed that the nuclei are similar but that the presence of obscuring material is a genuine difference in the two classes of objects. The role of orientation effects is then probably secondary and needs be studied with the help of complete samples.

2.5 Complete Samples

AGN and quasars come in a great variety of disguises that led to a complex classification. This is due in great part to the fact that some properties are enhanced when associated with a relativistic jet pointed towards us (and correspondingly weakened when the jet is pointed in other directions) while some other aspects may be hidden by obscuring material as noted in the previous subsection. It is worth noting that some of the differences are certainly intrinsic to the objects and not simply related to geometrical and/or Doppler boosting and/or absorption. Among those intrinsic differences is the presence of the blue bump and the existence of the jet in the first place.

In order to understand what are the intrinsic differences as opposed to viewing or orientation effects and in order to measure the effects related to viewing and orientation it is mandatory to obtain complete samples of AGN. This task is proving difficult because of the difficulty to find parameters that are indeed independent of orientation and absorption and that are unique signatures of AGN activity. The presence of hard X-ray radiation is for example a clear signature of AGN activity and, above the energy at which absorption is important, is rather insensitive to the presence of obscuration material. The X-ray emission of radio loud quasars is, however, more important than that of radio quiet quasars [19]. This may be due to the fact that a fraction of the X-ray emission is linked to relativistic jets. Thus samples based on hard X-ray flux measurements of AGN whilst unbiased towards obscured or unobscured AGN will contain biases related to the presence or not of jets and to their orientation with respect to the line of sight.

Polletta et al. [19] have computed the continuum spectral energy distribution of a small sample of quasars with very different radio properties for which far infrared data from ISO have been measured. This analysis which is based on a sample that is not complete suggests that the far infrared emission is the emission component that is least affected by biases. This may indeed be expected since dust emission occurs on large scales, is isotropic and is unrelated to the jet. Nonetheless if confirmed this implies that the dust properties of quasars are independent of their class, a non trivial result. Based on this preliminary result we suggest that, at least for quasars, complete samples should be based on far infrared measurements of objects that are identified by their X-ray emission to be AGN rather than starburst galaxies. This approach may not work in the case of BL Lac objects in which the far infrared emission is dominated by synchrotron emission.

2.6 AGN Versus Stars as Photon Sources in the Universe

Several authors (also in this volume) have noted the similar shapes of the star formation history in the Universe and of the QSO or AGN evolution. This similarity is then taken as a strong suggestion that star formation and AGN activity are intimately linked in the development of galaxies.

Dunlop [9] has compared the the radio luminosity of quasars as a function of redshift with the star formation history in the Universe and has come to the conclusion that there are 10^7 solar masses of stars formed per solar mass accreted into a black hole in a radio loud quasar. As far as energetics is concerned a 10 solar mass star is representative of a population and generates about $4 \cdot 10^{52}$ ergs over its lifetime (Maeder private communication). This means that for 10^7 solar masses some $4 \cdot 10^{58}$ ergs are generated. The energy liberated by the accretion of one solar mass assuming an efficiency of 10% is $2 \cdot 10^{53}$ ergs. It follows that the ratio of accreted energy generation onto massive black holes in radio loud quasars to that of nucleosynthesis in the Universe is 10^{-5}. In order to convert this ratio into a global estimate of the total rate of energy production by accretion onto all massive black holes to that due to nucleosynthesis it would be necessary to know the density ratio of AGN to the radio loud QSO's used in Dunlop (1998) and their respective average luminosities.

Fabian and Iwasawa [11] have used the hard X-ray background and generic AGN spectral energy distributions to estimate the total contribution of AGN to the photon flux and compared this with the photon flux of star formation. They conclude that the total power emitted by accretion is about 1/5 that due to stars in the Universe. A further comparison between star formation history and quasar evolution is presented in Franceschini et al. [12] and can also be used to derive that the ratio of accretion to star formation power is of the order of 10^{-1}.

The close ties between the star formation and AGN luminosity in the Universe as it evolved provides a fascinating example of the interconnection of phenomena which many of us would have thought independent.

2.7 What Makes an AGN Radio Loud?

A small fraction of AGN are radio loud. It would indeed be interesting to know this fraction with some precision as a function of the luminosity and redshift of the AGN. The origin of this difference cannot be sought in absorption as radio waves are undisturbed by neutral gas and dust, nor can it be sought in orientation and Doppler boosting effects, because radio galaxies show that a significant fraction of the radio luminosity is emitted isotropically. The presence or absence (or at least extreme weakness) of radio emission is therefore intrinsic to the objects.

Radio emission is related to the presence of jets in the AGN. The origin of the radio emission is therefore related to the presence of jets. The question may therefore be re-phrased to asking why a small fraction of AGN have jets and others not. The presence of magnetic fields and/or the rotation of the black hole certainly play a role in these question.

2.8 What Turns a Galaxy into an AGN?

Evidence for black holes in galaxies has now become strong, in particular in the center of our Galaxy [10]. The luminosity of the galactic center, and that of the nuclei of other galaxies in which black holes are probable, is, however, orders of magnitude less than the Eddington luminosities inferred from the masses of the black holes. The presence of the black hole is, therefore, not a sufficient condition for the presence of a bright AGN in a galaxy. There must be in addition to the black hole a sufficient quantity of matter that can be accreted in order to turn the central black hole of a galaxy into an active AGN. This matter can be in the form of diffuse gas accreted through a disk or in the form of a sufficiently dense cluster of stars surrounding the black hole.

2.9 What Holds the Broad Line Region Clouds Together?

Dietrich et al. [8] have recently shown by observing with high resolution and high signal to noise the broad lines of 3C 273 that the number of clouds is in excess of 10^8. This number is larger than the number of stars that have been suggested as a source of confinement for the clouds [21], [1], [2]. It is therefore probable that models in which the broad line region is seen as a continuous hydrodynamic medium will have to be preferred to those in which a collection of well defined clouds exist for a significant time. See Dietrich et al. [8] for a list of references to such models.

3 Multi-Wavelength Observations

All the questions discussed above have to be addressed with observations in several wavebands. This is even true for the question of the broad line cloud confinement in which a possible confining medium could have been detected through X-ray observations.

We will be in a very privileged position in the next years with respect to multi-waveband coordinated observations of AGN with the simultaneous presence in orbit of high energy observing satellites that will cover the energy domain from 10s of GeV (AGILE, AMS) to MeV and 100s of keV (INTEGRAL), to soft X-rays and ultraviolet (XMM, Chandra, Spektrum X-G).

These high energy instruments are complemented with a large array of very performing ground based telescopes in the visible, near infrared and radio domains of the spectrum and of the satellite SIRTF in the far infrared domain. We should thus be in a position to address many of the questions above in a very comprehensive way.

One can reasonably expect that well planned observations using all these facilities should allow us to solve the question of re-processing, that of the relationship between starburst and AGN activity and to obtain well defined complete samples in a convincing way. We should then be able to make very significant advances on many of the other topics mentioned here and in particular on the way in which material is accreted and the nature of the blue bump.

To be able to, and to have to, formulate the questions mentioned here is both a tribute to those who have pioneered the field of AGN, Lo Woltjer being prominent among them, and a sign that the subject is difficult and has been progressing only slowly once the bases were laid. It is our hope that the terms in which the questions are expressed and the tools that are now available will allow us to progress. We are well aware, though, that we are not the first to express these hopes and that observations of AGN have often provided new questions rather than answers to those they were expected to solve.

References

1. Alexander T. and Netzer H., 1994, MNRAS 270, 803
2. Alexander T. and Netzer H., 1997, MNRAS 284, 967
3. Antonucci R., 1993, Ann.Rev.Astron.Astrophys. 31, 473
4. Antonucci, R. and Miller J., 1985, Ap.J. 297, 621
5. Courvoisier T.J.-L. and Clavel J., 1991, A&A 248, 389
6. Courvoisier T.J.-L., Paltani S. and Walter R., 1996, A&A 308, L17
7. Delgado Gonzales R.M. and Heckman T., 1999, Ap&SS 266, 187
8. Dietrich M., Wagner S.J., Courvoisier T.J.-L., Bock H. and North P., 1999, A&A 351, 31
9. Dunlop J., 1998, in Observational Cosmology with the New Radio Survey, Dordrecht: Kluwer Academic Publishers, Astrophysics and space science library (ASSL) Series vol no: 226, 157
10. Eckhart A. and Genzel R., 2000, in Galaxies and their Constituents at the Highest Angular Resolution, International Astronomical Union. Symposium no. 205. Manchester, England, August 2000.
11. Fabian A.C. and Iwasawa K., 1999, MNRAS 303, L34
12. Franceschini A., Hasinger G., Takamitsu M. and Malquori D., 1999, MNRAS 310, L5
13. Haardt F., Maraschi L., and Ghisellini G., 1994, Ap.J. 432, L95
14. Koratkar A., and Blaes O., 1999, PASP 111, 1
15. Mushotzky R.F., Done C. and Pounds K.A., 1993, Ann.Rev.Astron.Astrophys. 31, 717

16. Oliva E.; Origlia L., Maiolino R. and Moorwood A.F.M., 1999, A&A 350, 9
17. Paltani, S., Courvoisier T.J.-L. and Walter R., 1998, A&A 340, 47
18. Paltani S. and Courvoisier T.J.-L., 1997 A&A 323, 717
19. Polletta M., Courvoisier T.J.-L., Wilkes B. and Hooper E.J., 2000, A&A in press
20. Pounds K.A., Nandra K., Stewart G.C., George I.M. and Fabian A.C., 1990, Nature 344, 132.
21. Scoville N. and Norman C., 1988, Ap.J. 332, 163
22. Shakura N. and Sunyaev R., 1973, A&A 24, 337
23. Shields G.A., 1978, Nature 272, 706
24. Torricelli-Campioni G., Foellmi C., Courvoisier T.J.-L., and Paltani S., 2000, A&A in press.

The Relationship Between Supernova Remnants and Neutron Stars

Franco Pacini

Arcetri Astrophysical Observatory and
Department of Astronomy and Space Sciences
University of Florence, Italy

Abstract. We comment briefly on facts and ideas concerning the relationship between neutron stars and supernova remnants and the role of the stellar magnetic field in determining the properties of these astrophysical objects.

1 Introduction

The interest for investigating the possible relationship between neutron stars and the activity observed in Supernova Remnants did develop in the years 1950-60's, when it was discovered that the continuous, amorphous, emission from the Crab Nebula over a large range of frequencies is due to the synchrotron process. Soon, it became clear that the persistence (more than 900 years after the original Supernova explosion) of radiation in the optical and X-ray bands requires an internal steady supply of relativistic electrons. Around the same time, the discovery of quasars and other violent events in galactic nuclei was posing a similar problem for the source of energy in these phenomena. This analogy was most clearly expressed by L. Woltjer during the Third Texas Symposium on Relativistic Astrophysics (1967) with the following statement:

"We require objects in the Q.S.O. that share the following with the Crab Nebula:

- *An electron energy spectrum corresponding to a spectral index 0.25.*
- *Large yield in relativistic electrons compared to all other modes.*
- *Burst-like activity with time scales less than a year.*
- *Anisotropic energy input leading to rather large-scale relativistic motions.*

In the case of the Crab Nebula the nature of the object is uncertain, but we have suggested (Woltjer, 1964) that the particle acceleration is related to the strong magnetic fields that may be met in objects near the Schwarzschild limit either static or collapsing. We wish to suggest that in the Q.S.O. the same process is going on a vastly larger scale and that the energy spectrum of the electrons is determined - in a yet unknown way - by the fundamental properties of this process"

The discovery of pulsars was about to be announced and much of the subsequent understanding of the activity in the Crab Nebula and active galactic nuclei has followed Woltjer's prescription. This explains why the organizers of this Conference - mostly devoted to extragalactic phenomena - have introduced in the program a subject where knowledge mostly stems from galactic observations.

In the following, I will comment briefly on facts and ideas concerning the relationship between neutron stars and the high energy activity in Supernova Remnants. For more details we refer the reader to broad reviews by Frail (1998) and Helfand (1998) and to a full coverage of the subject in the Proceedings of the Arcetri - Elba Workshop "Relationship between Neutron Stars and Supernova Remnants" (Bandiera et al., 1998).

2 The Crab Nebula and Other Supernova Remnants

Supernova remnants are usually classified in two extreme categories: shell-type and filled-center (plerions).

In the case of shell remnants, the edges of the source are bright, the interior is faint. The radio spectrum is steep ($S_\nu \propto \nu^\alpha$ with $\alpha \approx -0.5$) and it is produced by synchrotron radiation from relativistic electrons generated by shocks in the region where the expanding debris interact with the circumstellar/interstellar medium. Cas A is the prototype of shell-type remnants. Among the 215 catalogued Supernova remnants 85% are of this type (Green, 1996).

On the opposite side, the Crab Nebula is the typical plerion, where a central neutron star continuously converts its rotational energy into a magnetized relativistic wind. This wind expands and produces a center-filled nebular emission with a flat radiospectrum ($\alpha \approx -0.2$). It is not surprising that several remnants show both the central emission and the bright limbs since it is to be expected that some plerions would expand into a relatively dense medium (composite remnants).

In the case of the Crab Nebula, observations at various frequencies yield information about the properties of the spectrum, the morphology at various frequencies, the polarization, the dynamics. We can then use the theory of synchrotron radiation to deduce the properties of the magnetic field (strength, distribution, geometry), as well as the characteristics of the injected electrons. It turns out that the typical value of the field is around 3×10^{-4} gauss and the number of electrons injected is about $10^{40}s^{-1}$ (energy range $10^8 - 10^{13}eV$). The total nebular energy content (electrons and magnetic field) is of the order of 10^{49} ergs. We also recall that the "invisible" relativistic protons cannot be dominant over the electrons, as it happens for cosmic rays. Otherwise, the internal pressure would force the nebula to expand with an acceleration higher than observed.

Simple homogeneous models for the evolution of the Crab Nebula and similar sources powered by the loss of rotational energy of neutron stars have be computed (Setti and Woltjer, 1972; Pacini and Salvati, 1973; Bandiera et al., 1984). In these models the evolution of the spectrum and energy content (particles, fields) is determined by the pulsar input and by the adiabatic and radiative losses in the expanding bubble. Electrons are subject both to radiation and to adiabatic losses; the magnetic field suffers only adiabatic losses.

We need to know:

- the pulsar energy loss $L(t) = \frac{L_0}{(1+\frac{t}{\tau_0})^2}$ (constant until $t \simeq \tau_0$)
- the fraction of the energy input which goes into electrons and magnetic field.
- the size of the expanding bubble $R(t)$.
- the injected spectrum of the electrons $J(E) = KE^{-\gamma}$

In the Crab, the model can reasonably account for the strength of the nebular field $B \sim 3 \times 10^{-4}$ gauss, for the observed decrease of the radio flux 0.17 % yr^{-1} (Aller et al., 1986) and for the existence of an observed spectral break $\Delta\alpha_{r-opt} = 0.5$ at $\nu_b \sim 10^4$ GHz. The latter corresponds to the electron energy Eb at which synchrotron and adiabatic losses are equal. It also predicts $E_b \propto \frac{1}{B^2 t}$. Apart from the possibility of an intrinsic injection breaks, this is the only spectral break present for $t < \tau_0$. However, when $t > \tau_0$, another break appears in the spectrum, corresponding to the adiabatic evolution of $E_b(\tau_0)$.

Although this simple model explains the global characteristic of the Crab and some other sources, it has been noted that several plerions (e.g. 3C 58, G21.5 - 0.9; CTB 87) do not fit in the same scheme, because of the position of the break and/or because $\Delta\alpha \gg 0.5$. Particularly puzzling is the possible increase with time of the radio-flux from 3C 58 (Aller et al., 1986). These discrepancies indicate that not all plerions evolve in a Crab-like manner and that, perhaps, we have not yet fully understood the time evolution of the pulsar energy release (Woltjer et al., 1997).

Modelling becomes even more complex if one abandons the assumption of homogeneity and tries to understand the radial distribution of the nebular emission at various wavelengths or to compare the detailed maps which are now available at various frequencies for the Crab Nebula. These maps are beautiful but they are also a genuine nightmare for theorists and clearly require a level of understanding of the injection and transport of the nebular particles much higher than presently available. One is led to admit that we do not yet have an answer to some basic questions, such as: Are all relativistic electrons accelerated in the neutron stars magnetosphere? or, alternatively: Are the electrons accelerated / re-accelerated further out, in the nebular volume? How are they transported around, in the nebula? (see also, e.g. Amato et al., 2000).

3 Demography of Neutron Stars in SN Remnants and the Role of the Magnetic Field

In the past, it was widely held that plerions contain a neutron star while shell-type remnants do not, either because the explosion blows apart the entire star or because the central object becomes a black hole. This belief was largely based upon the lack of direct or indirect evidence for the presence of a compact object inside shell remnants (direct evidence would come from the observation of pulses, indirect from the detection of the relativistic wind).

There were, however, some inconsistencies. For instance, the estimate for the rate of core collapse Supernovae (roughly, one every 30-50 years) is about a factor two larger than the same estimate for the birth- rate of radiopulsars (roughly, one every 100 years). This did suggest the existence of a large fraction of neutron stars which do not appear as radiopulsars (Helfand and Becker 1984, Helfand 1998). Indeed, in recent years the observational evidence for the presence of neutron stars in Supernova remnants has changed, largely because of observations in the X and γ-rays bands. According to Frail (1998), at least 19 remnants contain neutron stars which manifest themselves in different ways. Seven of them are classical radiopulsars (only one third of the total!); 3 are X-ray binaries; 2 are slow X-ray pulsars; 2 are soft γ-ray repeaters; 5 are radioquiet neutron stars (detected through their thermal emission). Although this list is not updated, the picture has not changed substantially. The reasons which lead to different manifestations of neutron stars in Supernova remnants are not fully understood but one can make the following considerations.

The binary nature of the system (i.e. the role of accretion), the initial period of the neutron star and/or the magnetic field strength are likely to play an important role in determining the final result after the collapse. Probably these factors are not independent. In particular the initial rotation of neutron stars could be related to the strength of the coupling between the collapsing core and the stellar envelope (Tsuruta and Cameron 1966, Pacini 1983). In the past, it was generally assumed that all neutron stars would have magnetic fields $\leq 10^{12}$ gauss. If so, pulsars would remain detectable during the whole lifetime of the remnant (or longer). However, strong magnetic fields are likely to slow down the initial rotation. Even if the coupling is not effective during the collapse, the wind produced by a newly born neutron star with $B \gg 10^{12}$ gauss would damp the rotation very rapidly after the explosion. The stellar magnetic energy $\sim B^2 R^3$ could then become dominant over the rotational energy $\sim MR^2\Omega^2$ relatively soon after the birth of the neutron star ("magnetar"). This possibility was suggested, already before the discovery of pulsars, by L. Woltjer (1964). He suggested that some neutron stars could have magnetic fields of, say, $10^{14} - 10^{15}$ gauss and that their annihilation through flares could represent an important source of activity. More recently, this possibility has been revived as a possible explanation of the soft gamma-ray repeaters (Duncan and Thompson, 1992). Furthermore, the possible presence of a neutron star hidden in Cas A (with period $P \geq 0.7$ sec B $\geq 10^{14}$ gauss) was discussed by Cavaliere and Pacini (1970) and reiterated (Pacini, 1999) just before a central point source in Cas A was found by the Chandra X-ray satellite (Tananbaum et al., 1999). Observations in the various bands are currently underway and will, hopefully, provide more information about the nature and possible periodicity of this source.

Some neutron stars with fields around 10^{14} gauss have been tentatively identified in recent times on the basis of the slowing down rate. These determinations of the magnetic strength require a cautionary word. Indeed, one should remember that the standard estimate of the field value from the slowing down is based upon the energy loss in a dipole field. Any deviation from the dipolar geometry inside the speed of light cylinder (or, even worse, the existence of additional breaking mechanisms) would change the strength on the stellar surface. Quite generally, the release of magnetic energy through flares could perturb the field geometry and strength, thus becoming recognizable also by the variability of the slowing down rate and by a jittering of the braking index $n(\dot{\Omega} \propto \Omega^n)$ (note that this could also lead to erroneous estimates of the pulsar age!).

In conclusion, it is now evident that the lack of a radiopulsar or of the plerionic component cannot be taken as proof for the non-existence of a neutron star inside a Supernova remnant. Also, over the last 40 years or so, it has become clear that rotation and magnetic fields play a decisive role in the high energy phenomena observed around galactic collapsed objects. The same general ideas have been applied to explain the violent events occurring in active galactic nuclei, as indeed suggested by L. Woltjer in the statement reported at the beginning of this paper. Unfortunately we are still far from a complete observational knowledge of all aspects of these phenomena and, even more, from a full theoretical understanding.

Acknowledgements

I am indebted to E. Amato, R. Bandiera, M. Salvati, L. Woltjer for many discussions on the subject of this talk. This work was partly supported by grants from the Italian Space Agency and the Italian Ministry of Universities and Scientific/Technological Research.

References

1. Aller, H.D., Aller, M.F., Reynolds, S.P., 1986, BAAS 18, 1052.
2. Amato, E., Salvati, M., Bandiera, R, Pacini, F., Woltjer, L., 2000, A&A, 359, 1107.
3. Bandiera, R., Pacini, F., Salvati, M., 1984, Ap.J., 285, 134.

4. Bandiera, R., Masini, E., Pacini, F., Salvati, M., Woltjer, L., 1998, Proc. of the Workshop "The relationship between neutron stars and Supernova remnants", Mem. SAIt, vol. 69, n. 4.
5. Cavaliere, A., Pacini, F., 1970, Ap.J., 159, L21.
6. Duncan, R.C., Thompson, C., 1992, Ap.J. 392, L9.
7. Frail, D.A., 1998, in Proc. of the NATO ASI "The many faces of neutron stars", Kluwer Academic Publishers, pp. 179-194.
8. Green, D.A., 1996, A catalogue of galactic Supernova remnants, Mullard Radio Astronomy Observatory, Cambridge, UK.
9. Helfand, D.J., Becker, R.H., 1984, Nature 307, 215.
10. Helfand, D.J., 1998, in Proc. of the Workshop "The relationship between neutron stars and Supernova remnants", Mem. SAIt, vol. 69, n. 4, pp. 791-800
11. Pacini, F., Salvati, M., 1973, Ap.J., 186, 249.
12. Pacini, F., 1983, A&A, 126, L11.
13. Pacini, F., 1999, in Proc. of the IAU Symposium 195, Highly Energetic Physical Processes and Mechanisms of Emission from Astrophysical Plasmas, ASP Conferences Series (Martens and Tsuruta, eds.)
14. Setti, G., Woltjer, L., 1972, Ap.J. (Letters), 178, L17.
15. Tananbaum, B. et al., 1999, IAUC No 7246.
16. Tsuruta, S., Cameron, A.G.W., 1966, Nature, 211, 356.
17. Woltjer, L., 1964, Ap.J., 140, 1309.
18. Woltjer, L., 1967, Proc. Third Texas Symposium on Relativistic Astrophysics (unpublished preprint).
19. Woltjer, L., 1968, Ap.J., 152, L179.
20. Woltjer, L., Salvati, M., Pacini, F., Bandiera, R., 1997, A&A, 325, 295.

The Future of Ground-Based Optical Interferometry

Pierre Léna

Université Denis-Diderot (Paris VII)
& Observatoire de Paris, 92195 Meudon, France

Abstract. Optical interferometry in astronomy has demonstrated in the last decade its scientific productivity and is now reaching maturity. Numerous interferometers exist throughout the world, but the new century opens with giant telescopes ready to operate as optical interferometers of unprecedented sensitivity. Along with extrasolar planets and circumstellar environments, active galactic nuclei are one of the most important field where existing interferometers are expected to bring new results, able to reveal the 0.1-10 pc inner structure on tens of these objects. New proposals are emerging: new generation interferometers for ground-based interferometry ; coherent combination of the Mauna Kea (Hawaii) telescopes using optical fibers coupling (OHANA project); extension of the sub-millimetric ALMA array to the mid-infrared. All these projects would have a direct impact on active galactic nuclei studies. Is the future of optical interferometry on the ground or in space, where several missions are in preparation?

1 Introduction

Although the emblematic figure of Albert Michelson (1852-1931) has been decisive with the first measurement of a stellar diameter, it is worth recalling here, in the happy circumstance of L. Woltjer's anniversary celebration, that since two centuries Europe has played an intense role in the development of high angular resolution astronomy, as summarized in Table 1 (see also the historical compendium of fundamental papers in [1], [2]). In the 1970s, speckle interferometry led to a deep understanding of the astronomical seeing and to related properties of the Earth atmosphere in image degradation. This opened the way to adaptive optics [3], which has matured extremely fast in the 1990s: it now equips all large ground based telescopes and produces first rank science [4] with the help of deconvolution techniques for image restoration. After a long eclipse, optical interferometry steadily progressed since 1975 at visible then near-infrared wavelengths: it has led to numerous small or medium size instruments exploring its basic concepts and beginning scientific production. These are opening the way to larger instruments, on the ground or in space, since adaptive optics makes now possible an efficient coupling of large ground based telescopes in an interferometric mode.

Several questions are now open for the future: what will be the ultimate performances of ground-based interferometers, in terms of sensitivity and astrometric accuracy? will ground-based arrays be in competition with single pupil giant telescopes of decametric or even hectometric sizes? will instead space instruments be the only ones to be seriously considered?

Table 1. Europe and High angular resolution in astronomy

1802	Interference fringes and wave nature of light	T. Young, London
1868	Method of stellar interferometry	H. Fizeau, Paris
1873	Upper limit of stellar diameters	E. Stéphan, Marseille
1950	First radio-interferometer	M. Ryle, Cambridge
1970	Speckle interferometry	A. Labeyrie, Paris
1975	Coupling of optical telescopes	A. Labeyrie, Paris
1976	Speckle imaging by triple correlation	G. Weigelt, Nuernberg
1987	*Very Large Telescope Interferometer* decision	ESO[†]
1989	First adaptive optics astronomical image	G. Rousset *et al*, Paris
1996	First interferometric image in the optical	J. Baldwin, Cambridge
2001	Expected VLTI first light	ESO[†]

[†] European Southern Observatory

2 Scientific Results from Operating Interferometers

Table 2 gives the list of existing or planned optical interferometers and their main characteristics. Descriptions of every instrument, with its performances and specific references, could be found in the recent (2000) Symposium *Interferometry in Optical Astronomy* [7]. Most of the instruments belonging to the first category (small or medium size, i.e. telescopes of less than 1.5 m) are operating and have been producing scientific data during the 1990s, while the interferometers made with 8-10 m telescopes will only become operational in 2001 or beyond. Space interferometers are in the design phase for launch during the mid- or late 2000 decade. Finally, some proposals of new interferometric configurations or new instruments, to be discussed below, are currently under review or partial implementation.

The 1990 decade saw a steady increase of scientific production by optical interferometers at wavelengths ranging from 500 nm to 3.5 μm. Table 3 clearly demonstrates that this production has been exclusively dealing with the (relatively) easiest, namely observations of stars or stellar environments: no result yet has been obtained on extragalactic sources. This is clearly a consequence of the limiting magnitudes imposed by either the small diameter of

Table 2. Main optical interferometers in the world (2000)

Name	Number of telescopes	Diameter (m)	Maximum baseline (m)	Date of operation	λ	Location
Small- or medium-size ground-based interferometers[a]						
CHARA	6	1	350	2000	vis-K	Mt. Wilson, California
COAST	5	0.4	22	1992	J-K	Cambridge, UK
GI2T	2	1.5	65	1990	vis	Calern, France
IOTA	3	0.4	38	1995	H-K	Mt. Hopkins, Arizona
MIRA-I	2	0.25	4	2001 (?)	K	Tokyo, Japan
NPOI	3	0.5	250	1998	vis	Flagstaff, Arizona
PTI	3	0.4	110	1995	H-K	Mt. Palomar, California
SUSI	2	0.14	640	1990	vis	Narrabri, Australia
Large ground-based interferometers						
VLTI	4 + 3	8.2 + 1.8	130 to 202	2001-2003	J-N	Cerro Paranal, Chile
Keck-I	2 + 4	10 + 1.8	85 to 135	2001	J-N	Mauna Kea, Hawaii
LBT[b]	2	8.4	23	2002	J-K	Mt. Graham, Arizona
Proposed future ground-based interferometers						
OHANA[c]	≤ 7	3.6 to 10	800	?	vis-K	Mauna Kea, Hawaii
ALMA[d]	≤ 4	≤	10^4	?	≥ 10μm	Chajnantor, Chile
IRVLA[f]	27	3	1000	?	J-K	Mauna Kea, Hawaii
OVLA[g]	27	1.5	≤ 10^4	?	vis	?
Space interferometers						
SIM	2	0.3	10	2006	vis	Heliocentric Earth trailing
IRSI (Darwin)	5	1	10-20	ca. 2010	10 μm	ca. 1 a.u. from Sun
TPF[h]	4	3.5	75 to 1000	≥ 2008	3-30 μm	L2 or Earth drift-away

[a] This list does not include the *Infrared Spatial Interferometer* (ISI) of Mt. Wilson, which is based on heterodyne detection at 10.6μm.
[b] With both telescopes on the same mount, LBT is the only ground-based interferometer with a large field-of-view (ca. 1 arcmin).
[c] Proposed fiber-linked combination of existing large telescopes on Mauna Kea summit. See text.
[d] Proposed addition to the millimetric ALMA array for mid-infrared (10 μm and above) observations. See text.
[f] Proposed new array for Mauna Kea. See text.
[g] Proposed array using pupil densification for exo-solar planets imaging.
[h] New NASA mission concepts are introduced in 2000, such as SPIRIT and SPECS. See text.

the telescopes, or the lack of adaptive optics to improve the sensitivity, or mediocre transmission and/or detectors of limited performances in prototype instruments. This situation will soon change. Nevertheless these programs have led to some remarkable achievements and solved most of the practical problems of interferometers operation.

- Adaptive optics, which is quite mandatory for interferometry as soon as the telescope size is significantly larger than the atmospheric coherence length r_o, is now operational on many telescopes with natural reference stars as faint as $m_V = 14 - 16$, allowing a wide range of interferometric science.

Table 3. Optical interferometry refereed scientific publications[a]

Topic	Number of papers
Stellar angular diameters	31
Shells of late type stars	15
Be/P Cyg shells	12
Binary stars orbits	17
Stellar morphology & atmospheres	10
Wide angle astrometry	2
Novae	1
Cepheids	1
Young stellar objects	2

[a] During the decade 1990-1999. After [5]

- Absolute accuracy on visibility modulus (fringe contrast) has improved by almost an order a magnitude with the reduction of atmospheric coherence losses using optical fibers as filters: it now reaches 0.3 % and may still improve. As a consequence, even a fairly limited number of visibility values (i.e. baselines) may strongly constrain the model used to represent the source. A beautiful demonstration of this is given by the first direct detection of photospheric pulsations of the Mira star R Leonis with the IOTA interferometer, equipped with the FLUOR instrument [8]. A similar work carried on the semi-regular variable SW Vir in the K band is able to detect photospheric diameter variations of 3 %.
- The constraints put on a model by a very limited number of visibility values are well illustrated by the detection of the accretion disc of the star FU Ori, observed at the Palomar Tesbed Interferometer (PTI) and at IOTA with up to 4 milliarcsec resolution (2 a.u.) in H and K bands [9]. Even with visibility accuracies of 5 %, the disc thermal emission is clearly resolved, leading to a deduced inner radius of 13 $R_\odot$ and an accretion rate of $8\times10^{-5} M_\odot$/yr. This particular result is especially encouraging for future observations of AGNs accretion discs.
- Image reconstruction by using phase closure with more than one telescope pair has progressed: the COAST interferometer shows the way with images of the symbiotic star CH Cygni at 905 nm obtained with five telescopes (nine baselines) [10]: phase errors are reduced to as little as 2°. The prospect for complete image reconstruction is shown by the remarkable work carried in the K-band with the Keck I 10-m telescope and a pupil mask of 21 apertures disposed over an image of the primary. In this case a complex interferogram is obtained, containing the inter-

ference of all the pairs of apertures: a similar interferogram would result from the simultaneous use of a number of independent telescopes. Reconstructed images of the Wolf Rayet star WR104, at different epochs, show the ejection of matter by the star and the formation of a spectacular, rotating "pinwheel nebula" [11].
- After a long period where interferometric observations were limited to the visible and near-infrared (except at 10 μm with the heterodyne interferometer ISI), the demonstration of wide-band interferometric operation in the thermal infrared has been given in the L band (3.5 μm), again with the IOTA/FLUOR interferometer [12]. The feasibility of operation at longer wavelengths, where the thermal background considerably increases, has not yet been proven but instruments for the VLT and Keck interferometers are designed for 10μm operation (see below) and there is no reason to expect fundamental difficulties [13].
- Spectral interferometric imaging, with resolutions as high as 10^3 in the J and K bands or 3×10^4 in the 570-700 nm range is now obtained, uniquely for the moment with the GI2T interferometer, using dispersed fringes in a short exposure mode. Remarkable results have been obtained on the star R Cas in the infrared [14] and on the equatorial disc of the binary star ζ Tau [15]. Future interferometric instruments will extensively use this spectro-interferometric mode of observation.
- Astrometry with optical interferometers on the ground is based on the differential method demonstrated by Shao & Colavita [16]: differential piston noise, due to the atmosphere, affects the relative position of two nearby (within ca. 1 arcmin) stars simultaneously observed and sets the ultimate reachable accuracy. The rms value of this noise is given by

$$\sigma_{\delta_2 - \delta_1} \approx 300 B^{-2/3} \theta T^{-1/2} \text{ arcsec}$$

where δ_1 and δ_2 are the respective atmospheric pistons on stars 1 and 2, θ their angular distance, B the baseline and T the integration time. This expression favors the long bases and the long exposures. For $B = 100$ m, $\theta = 10''$ and $T = 30$ mn, one gets $\sigma_{\delta_2 - \delta_1} \approx 16$ microarcsec, a very small value indeed. Differential interferometric astrometry is applicable to relative motion of nearby objects (binaries) or to the the detection of the reflex motion of a star having one or several planetary companions: great future developments in this direction are expected with dedicated astrometric interferometers such as NPOI. The Palomar Testbed Interferometer (PTI) has currently demonstrated a differential astrometric night-to-night repeatability of 100 μarcsec, reaching 170 μarcsec on the bright star 61 Cyg over a 70 day time scale [17].

All these results show the progressive mastering of optical interferometry and have laid the foundations for the more ambitious interferometers to come.

3 Interferometry with 8-10 m Telescopes

Three interferometers belong to this category: VLT, Keck and LBT. The latter is original in three respects: it has a rather short baseline (23 m), it is fully steerable (altaz mount) and as a consequence it has no long delay line and a unique interferometric field-of-view of ca. 1 arcmin. It is impossible here to give an extensive view of all the observing modes and scientific objectives of these three instruments: let us briefly examine their potential for AGNs studies.

For visibility amplitude measurements, the sensitivity of an interferometer is simply given by the expression of the signal

Signal = Source Intensity × Telescopes area × Spectral bandwidth
× Exposure time × Visibility amplitude × Instrument transmission
× Strehl ratio

while the noise is given as usual by detector (J,H,K bands), background (L band and beyond) or signal (visible) noise depending of the circumstances. This expression favors large telescopes, under the condition of adaptive optics availability. Strehl ratios above 0.5 are expected above a wavelength of $1\mu m$ for the 8-10 m telescopes. The exposure time is the most critical item: it clearly separates two regimes, known in interferometrists jargon as "bright" and "faint". In the former, exposure times are limited to the characteristic time of the atmospheric piston, i.e. some 50-100 ms, with a strong impact on the sensitivity. In the latter, long exposure times (minutes to hours) can be obtained by setting the adaptive optics and tracking the atmospheric piston on a "bright" star located close ($\leq$ 1-2 arcmin) to the "faint" target (as above in differential astrometry). The long time integration allows to measure visibilities on very faint objects, especially extragalactic ones (AGNs). This *dual feed* mode will equip the VLTI and the K2I.

To be more specific on sensitivities, let consider the VLTI case. The VLTI will be equipped with a first light test instrument (VINCI), a near-infrared instrument (AMBER) able to accommodate 3 (expandable to 4) telescope inputs [18] and a 10 μm instrument with two inputs (MIDI), see [19]. In addition a service instrumentation (PRIMA) will provide a fringe tracker (H band) for dual feed operation and an accurate internal metrology for astrometry. Table 4 and 5 provide the sensitivity in "bright" and "faint" modes, in terms of signal-to-noise ratio on visibility amplitudes. Similar numbers are expected for the Keck Interferometer.

A comparison of MIDI sensitivities (Table 4) with a list of AGNs observable from Chile (Table 6) shows their possible observation in the "bright" mode. The maximum angular resolution of the VLTI (UTs) at 10 μm is 20 mas which corresponds to a linear size of 1.8 pc for the close-by Seyfert galaxy NGC 1068, while for more distant objects it increases to 5 pc. These resolutions give access to the outer shape of the dust tori and possibly to the inner hole for the closest objects. AMBER observations can give a five-

Table 4. The VLTI MIDI Instrument sensitivity

Spectral resolution $\lambda/\Delta\lambda$	N limiting magnitude
No fringe dispersion	5[a]
50	2.7
500	1

Assumptions are: 100 ms exposure ("bright" object), Signal-to-noise ratio of 10, Visibility amplitude =1. If fringe tracking is done on a "bright" object, magnitude gain for the target detection is 2.5 for a 10 s integration time. Data from [19].
[a] or 400 mJy.

Table 5. The VLTI AMBER Instrument sensitivity

Mode	Exposure time	Spectral resolution	Limiting K-magnitude (UTs[a])	Limiting K-magnitude (ATs[a])
High precision	0.01 s	5	11.3	8.0
High sensitivity	0.1 s	5	13.2	9.9
Fringe tracking[b]	4 hr	5	17	12.1
Fringe tracking[b]	4 hr	100	16.5	11.6
Fringe tracking[b]	4 hr	1000	15.0	11.1
Fringe tracking[b]	4 hr	10^4	14.2	9.6

Assumptions are: 100 ms exposure ("bright" object), Signal-to-noise ratio of 5. Visibility amplitude = 1. Strehl ratio of 0.5. One polarization is used. Data from [18].
[a] UTs: 8.2 m telescopes; ATs = auxiliary movable 1.8 m telescopes.
[b] The fringe tracking is made on a "bright" source close to target.

to ten-fold (at 1 μm) angular resolution gain and its sensitivity for "bright" objects ($m_K \approx 13.2$ for UTs) gives in principle access to numerous Seyfert and ultraluminous IR galaxies (ULIRG): the "high sensitivity" mode is well adapted, with visibility accuracies of 1 %. Yet, one question to be solved is the fraction of the luminosity which is contained in a region of size comparable to the diffraction-limited field-of-view of the individual telescopes. Adaptive optics imaging with 8-10 m telescopes will bring the answer and provide final information for interferometric observation of AGNs. Of special interest here is the capability of the LBT interferometer, with an interferometric field-of-view larger than the diffraction figure of an 8 m pupil: it will be able to provide maps of AGNs nuclei at 23 m baseline helpful to guide further investigation with the VLTI larger baselines (see Fig. 1).

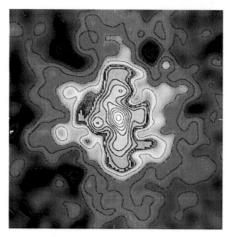

Fig. 1. The nucleus of NGC 1068, imaged at 2.2µm with the 3.6-m CFHT and adaptive optics (PUEO). Image size is 2.2″ × 2.2″ and resolution 0.11″ [21]

The VLTI capabilities in imaging, spectroscopy and astrometry at 1-2 µm with AMBER are well suited for AGNs studies. In addition to the main telescopes baselines, imaging is provided by the array of 30 auxiliary telescopes (AT) positions over an area 120 m (E-W) × 200 m (N-S). Spectroscopy with AMBER (Table 5) can provide high resolution for any "bright" object or when reference is available. Finally, the astrometric sensitivity of PRIMA, reaching 10 µarcsec, will allow detection of proper motions within the nucleus when a reference background star ($m_K \leq 15 - 16$) is available [20]: a 1000 km/s jet at 10 Mpc gives a displacement of 20 µarcsec/year.

4 Future Ground-Based Interferometers

The VLT or Keck configurations are a compromise between a classical telescope and an interferometer. A dedicated powerful interferometer would have: large telescopes (≥ 3 m) or conversely a large number of smaller ones phased in either case with adaptive optics of high performance and movable mounts allowing array reconfiguration (as in the VLA); many apertures for imaging capability; long (kilometric) baselines for adequate resolution on exo-solar planets and AGNs. These goals represent some technical and cost difficulties which have up to now postponed any decision on such programs, although some have been proposed [5]. The extreme dilution of the entrance pupil in such imaging arrays makes the fringe analysis difficult, and the Labeyrie's

Table 6. Extragalactic sources for VLTI/MIDI observation

Object	10 μm Flux density (mJy)
NGC 253	2900
NGC 1068	780
NGC 1808	680
NGC 2377	440
NGC 2818	440
NGC 3758	4250
NGC 3783	440
NGC 4594	570
ESO 445-G50	770
Mark 463	600
NGC 7469	600
NGC 7582	724

Data from Leinert C. et al., Preliminary MIDI Report, 1998.

proposal of pupil densification [22] is a very promising one to solve this problem, at the expense of the field-of-view: pupil densification allows to concentrate the light in a single PSF core in the case of a highly diluted array.

While waiting for the construction of such new generation interferometers, intermediate ground-based programs may be appropriate to increase the available angular resolution (OHANA) and to access the mid-infrared domain (ALMA extension): both actions are of limited cost and seem well suited for AGNs studies.

The OHANA (Optical Hawaiian Array for Nanoradian Astronomy) proposal [23] aims to progressively use the Mauna Kea summit 3.6 to 10 m telescopes (up to 7) as elements of a 800 × 300 m array, taking advantage of existing adaptive optics on these telescopes and of the demonstrated possibility (by the IOTA/FLUOR interferometer) of a fiber-linked interferometric operation in the range 500 nm - 2.5 μm. Figure 2 shows the potential baseline configurations. Nearly diffraction-limited images (Strehl ratio $\geq$ 0.3) are extracted from each telescope with single-mode optical fibers, which transport the beam in existing underground ducts to a common recombination area where a delay line is installed in air or vacuum. An angular resolution of 0.25 mas at 1 μm plus some imaging capability, with a sensitivity ("bright" sources) of $m_K \approx 14 \pm 1$ becomes perfectly suitable for AGNs central cores:

the Broad Line Region becomes accessible on objects as far as 100 Mpc, while the 0.02 pc resolution at 10 Mpc may resolve the accretion disc on a few objects such as NGC 1068. After joint discussions between the various Observatories which may become involved, a test program aiming to check fiber injection and light transport, then to couple a pair of telescopes (probably CFHT and Gemini, or CFHT and Keck) has been engaged in 2000.

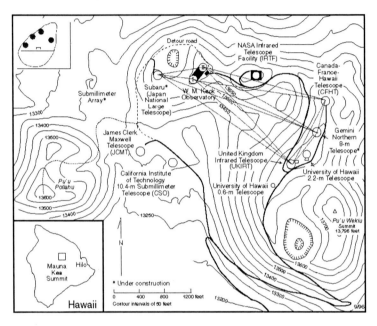

Fig. 2. The Mauna Kea summit and the potential OHANA baselines. The VLTI configuration and baselines are in the insert at the same scale.

The ALMA project is a millimetric and sub-millimetric large array, comprising at least 64 individual 12-m telescopes, to operate at 5100 m on the Chilean-Argentina border in the frequency range 70-900 GHz (i.e. down to 350 μm). The exceptional atmospheric conditions of the site will open relatively good far infrared windows [24] which could be observed nowhere else on Earth (except possibly the South Pole). The 10 km extension of the plateau would give a potential angular resolution of 0.2 mas at 10 μm, just matching the OHANA resolution at 1 μm. As the optical quality of the planned submillimetric telescopes is not adequate for short wavelengths, alternate solutions should be explored to install at least a few antennae capable of short wavelengths operation. The original antenna design calls for aluminium plates with grooves enabling solar observations: indeed, some plates should be polished to allow mid-IR operation. The aimed surface accuracy is 25 μm rms

Science and Technology of a 100-m Telescope: The OWL Concept

Roberto Gilmozzi, Philippe Dierickx, and Guy Monnet

European Southern Observatory
Garching bei München, Germany

Abstract. ESO is developing the concept of a ground-based, 100-m class optical telescope with segmented large mirrors and an integrated diffraction limit capability, christened OWL for its keen night vision. At ten times the collecting area of every telescope ever built put together, OWL will have limiting magnitudes of 37-38, angular resolutions of 1-2 milliarcseconds, and a price tag that does *not* follow the historical $D^{2.6}$ cost law. This paper describes the design progress and discusses some of the possible science cases, including the determination of H [not Ho] unencumbered by local effects, the study of every SN ever exploded at any $z < 10$, the spectroscopy of extra-solar planets, studies of ultrahigh frequency phenomena, imaging of stellar surfaces.

1 Introduction

Since 1997, ESO is developing the concept of a ground-based 100-m class optical telescope. It has been christened OWL for its keen night vision: this also stands for *OverWhelmingly Large* (showing either the *hubris* of astronomers or their distorted sense of humor). The starting point was the assessment that the next major step to unravel the structure and evolution of our Universe after HST and the 8-10-m class Keck/VLT generation requires an order of magnitude increase in aperture size. A similar assessment was made in the U.S. in 1996. The challenge and the science potential are formidable – and highly stimulating: a 100-m diffraction-limited optical telescope would offer 40 times the collecting power of the whole VLT with the milliarcsecond imaging resolution of the VLTI.

A number of Extremely Large Telescopes (ELTs) concepts have been proposed, some close to actual funding like the 30-m California Extremely Large Telescope (CELT). Reaching this mammoth sizes with full diffraction-limited performance, a global cost that cannot realistically exceed a billion Euros and a reasonable (≤ 12 years) time frame is the challenge that we are set to address in our present, 3 years long, phase A study. Its primary objective is to *demonstrate feasibility within proven technologies*, but provision is made for likely technological progress allowing either cost reduction or performance improvement, or both.

A few principles, mostly borrowed from recent developments in the art of telescope making, hold the key to meet these harsh requirements: optical

19. Leinert, C. et al. (2000) 10-μm interferometry with the VLTI MIDI instrument: a preview *in* Interferometry in Optical Astronomy, Léna, P. & Quirrenbach, A. (Ed.) Proceedings of SPIE, **4006**, 43-53
20. Delplancke, F., Lévêque, S., Kervella, P., Glindemann, A., d'Arcio, L. (2000) Phase-referenced imaging and μarcsec astrometry with the VLTI *in* Interferometry in Optical Astronomy, Léna, P. & Quirrenbach, A. (Ed.) Proceedings of SPIE, **4006**, 365-376
21. Rouan, D., Rigaut, F., Alloin, D., Doyon, R., Lai, O., Crampton, D., Gendron, E., Arsenault, R. (1998) Near IR images of the torus and micro-spiral structure in NGC 1068 using adaptive optics, Astron.Astrophys. **339**, 687
22. Labeyrie, A. (1996) Resolved imaging of extra-solar planets with future 10-100 km optical interferometric arrays, Astron.Astrophys.Suppl.Ser. **118**, 517-524
23. Perrin, G., Lai, O., Léna, P., Coudé du Foresto, V. (2000) A fibered large interferometer on top of Mauna Kea *in* Interferometry in Optical Astronomy, Léna, P. & Quirrenbach, A. (Ed.) Proceedings of SPIE, **4006**, 708-714
24. Paine, S., Blundell, R., Cosmo Papa, D., Barrett, J.W. (2000) A Fourier Transform Spectrometer for measurement of atmospheric transmission at submillimetric wavelengths, Publ.Astr.Soc. Pac. **112**, 108-118
25. Schneider, J., Coudé du Foresto, V., Léna, P. (2000) Mid-infrared interferometry at the ALMA site, in preparation
26. Unwin, S.C., Shao, M. (2000) The Space Interferometry Mission *in* Interferometry in Optical Astronomy, Léna, P. & Quirrenbach, A. (Ed.) Proceedings of SPIE, **4006**, 754-761
27. Röttgering, H., Granato, G.L., Guiderdoni, B., Rudnick, G. (2000) Scientific potential of infrared interferometry from space *in* Interferometry in Optical Astronomy, Léna, P. & Quirrenbach, A. (Ed.) Proceedings of SPIE, **4006**, 742-753

2. Lawson, P.R. (2000) *in* Principles of Long Baseline Stellar Interferometry, 1999 Michelson Summer School, Jet Propulsion Laboratory
3. Roddier, F. (Ed.) (1999) Adaptive Optics in Astronomy, Cambridge University Press
4. Bonnacini, D. (Ed.) (1999) Astronomy with Adaptive Optics: Present Results and Future Programs, European Southern Observatory, Garching
5. Ridgway, S.T., Roddier, F. (2000) An Infrared Very Large Array for the 21st Century *in* Interferometry in Optical Astronomy, Léna, P. & Quirrenbach, A. (Ed.) Proceedings of SPIE, **4006**, 940-950
6. Saha, S.K., Emerging trends of optical interferometry in astronomy (1999) Bull.Astron.Soc.India **27**, 441-546
7. Léna, P., Quirrenbach, A. (Ed.) (2000) Interferometry in Optical Astronomy, Proceedings of SPIE, **4006**
8. Perrin, G., Coudé du Foresto, V., Ridgway, S.T., Mennesson, B., Ruilier, C., Mariotti, J.-M., Traub, W.A., Lacasse, M.G. (1999) Interferometric observations of R Leonis in the K band, Astron.Astrophys. **345**, 221-232
9. Berger, J.-P., Malbet, F., Colavita, M.M., Ségransan, D., Millan-Gabet, R., Traub, W.A. (2000) New insights in the nature of the circumstellar environment of FU Ori, *in* Interferometry in Optical Astronomy, Léna, P. & Quirrenbach, A. (Ed.) Proceedings of SPIE, **4006**, 597-604
10. Young, J.S., Baldwin, J.E., Boysen, R.C., George, A.V., Haniff, C.A., Mackay, C.D., Pearson, D., Rogers, J., Warner, P.J., Wilson, D.M., Wilson, R.W. (2000) Recent astronomical results from COAST, *in* Interferometry in Optical Astronomy, Léna, P. & Quirrenbach, A. (Ed.) Proceedings of SPIE, **4006**, 472-480
11. Tuthill, P.G., Monnier, J.D., Danchi, W.C. (1999),A dusty pinwheel nebula around the main star WR104, Nature, **398**, 487
12. Mennesson, B., Coudé du Foresto, V., Perrin, G., Ridgway, S.T., Ruilier, C., Traub, W.A., Carleton, N.P., Lacasse, M.G., Mazé, G. (1999) Thermal infrared stellar interferometry using single mode guided optics: first results, Astron. Astrophys. **346**, 181-189
13. Porro, I., Graser, U., Leinert, C., Lopez, B., von der Lühe, O., (2000) Estimated performance for 10 μm interferometry at the VLTI with the MIDI instrument, Publ.Astr.Soc.Pac., submitted
14. Weigelt, G., Mourard, D., Abe, L., Beckmann, U., Chesnbeau, O., Hillemans, C., Hoffmann, K.-H., Ragland, S., Schertl, D., Scholz, M., Stee, P., Thureau, N., Vakili, F. (2000) GI2T/REGAIN spectro-interferometry with a new infrared beam combiner *in* Interferometry in Optical Astronomy, Léna, P. & Quirrenbach, A. (Ed.) Proceedings of SPIE, **4006**, 617-626
15. Vakili, F., Mourard, D., Stee, P., Bonneau, D., Bério, P., Chesneau, O., Thureau, N., Morand, F., Labeyrie, A., Tallon-Bosc, I. (1998) Evidence for one-armed oscillations in the equatorial disc of ζ Tau from GI2T spectrally resolved interferometry, Astron. Astrophys. **335**, 261-265
16. Shao, M., Colavita, M.M. (1992) Potential of long baseline infrared interferometry for narrow-angle astrometry, Astron. Astrophys. **262**, 353-358
17. Lane, B.F., Colavita, M.M., Boden, A.F., Lawson, P.R., (2000) Palomar Testbed Interferometer - Update *in* Interferometry in Optical Astronomy, Léna, P. & Quirrenbach, A. (Ed.) Proceedings of SPIE, **4006**, 452-458
18. Petrov, R. et al. (2000) AMBER: the near infrared focal instrument for the VLTI *in* Interferometry in Optical Astronomy, Léna, P. & Quirrenbach, A. (Ed.) Proceedings of SPIE, **4006**, 68-79

which means that quasi-static aberrations could be reduced to an acceptable level by low-order active optics. Under good seeing, the 12-m telescopes shall be close to diffraction-limited regime and only a tip-tilt correction is necessary. Light could be fed at the focus in a fiber and carried to the recombination area through mid-infrared fibers (a rapidly progressing technology still requesting developments). Wavelength range, angular resolution and sensitivity are strong arguments to explore a modest extension of ALMA in the mid-infrared, using the whole infrastructure without high additional costs [25]. Beside AGN studies, such a complement to the existing ALMA project would be of high value for the study of extra-solar planets, since the linear resolution at 10 pc is 1 $R_\odot$ and is amply sufficient to separate even the closest "hot Jupiters" from their parent star.

5 Conclusion

Adaptive optics, eventually coupled with laser reference star, has allowed to break the seeing barrier and to construct large size optical telescopes on the ground with an efficiency comparable to space instruments. No such panacea yet exists for ground-based interferometers, the sensitivity of which remains fixed by the effect of atmospheric piston: it limits the exposure time unless a nearby reference source can be found for fringe tracking. In addition, piston noise limits the ultimate astrometric accuracy to some 10 μarcsec for hectometric baselines.

To illustrate this point, it is worth noticing that the *United States Astronomy and Astrophysics Survey Committee*, in the preliminary version of its Report for the decade 2000, did not retain an interferometric project on the ground but recommends "... a single aperture far infrared telescope SAFIR ... and then ultimately in the decade 2010-2020 build on the SAFIR, TPF and NGST experience to assemble a space-based far infrared interferometer".

The astrometric SIM mission will reach relative narrow-field accuracy of 1 μarcsec in a 1-hour measurement and an accuracy of 4 μarcsec on global astrometry after a 5 years mission [26]. It will be followed by the Terrestrial Planet Finder mission, possibly merged with the European InfraRed Space Interferometer (IRSI-Darwin) mission: both being infrared interferometers aimed mainly at extra-solar planets detection but also capable of AGNs studies, imaging and spectroscopy [27].

Yet, even if space interferometry will ultimately dominate and and thanks to the considerable efforts made since 1975, there are broad perspectives for existing and future ground-based interferometers on relatively bright objects (m_H or $m_K \leq 15 - 16$).

References

1. Lawson, P.R. (Ed.) (1997) Selected Papers on Long Baseline Stellar Interferometry,SPIE Milestone Series, Vol. MS139

segmentation as pioneered by the Keck, massive production of standardized mirrors from the Hobby-Eberly, active optics control and system aspects from the VLT. The most critical aspect is to develop the means to reach diffraction-limited images in "large" fields (a few arcminutes in the near-IR). This is the goal of the so-called multi-conjugate adaptive optics concept, whose principle has recently been confirmed experimentally. Tremendous pressure is building up to implement such a capability into existing large telescopes, and rapid progress in the underlying technologies is taking place, *e.g.* to fabricate low-cost deformable mirrors with tens of thousands actuators from integrated circuits techniques.

1.1 Why 100-m? Starting with the result that $\sim$ 100-m diameter is needed to achieve spectroscopy of the faintest objects imaged by the NGST, a deeper analysis of key science cases shows the need for gains in depth ($\geq 35^{\text{th}}$ magnitude) and spatial resolution (1 milliarcsecond) to really achieve qualitative and quantitative scientific improvements. Only a large telescope of this size could offer it.

1.2 Can we afford it? Succeeding with OWL means breaking time-honored trends like doubling the diameter D only every 30 years, with a $D^{2.6}$ cost law and primary mirror production speed of 1m linear per year. An especially encouraging case concerns the primary (and secondary) mirror fabrication: industry indicated a baseline of 6-8 years for producing the full 100-m mosaic at affordable cost. One of our main challenges is to achieve such a huge reduction in time and cost for *all* major components of the Telescope. There are vocal supporters for a more "conservative" approach in any Extremely Large Telescopes Meeting; the burden of the proof is clearly in our camp and a major item for our Concept Study (to be completed early 2003).

1.3 Resolution, resolutely. Angular resolution and sensitivity are the highest priority requirements. Figure 1 illustrates the effect of increased resolution by showing the same hypothetical 0.6 × 0.6 arc seconds field, as seen by a seeing-limited telescope under best conditions ($\sim$ 0.2 arc sec), HST, an 8-m diffraction-limited telescope and OWL. There are also technical reasons for striving for high spatial resolution. Instruments designed for a seeing-limited 100-m would be impossibly large or wholly inefficient or both. There are exceptions, e.g. a mosaic of multi-band imagers, but none covering core OWL science cases. The price to pay to get efficient instruments actually smaller than their present VLT counterparts is of course the exacting adaptive optics system which must be incorporated in the overall concept right from the start.

1.4 OWL's performance. At ten times the combined collecting area of every telescope ever built, a 100-m filled aperture telescope would open new horizons in observational astronomy – going from 10m to 100m represents a "quantum" jump similar to that from the naked eye to Galileo's telescope.

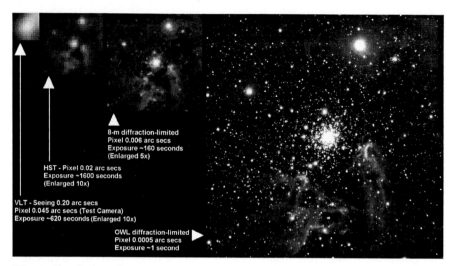

Fig. 1. Resolution, from 0.2 arc seconds seeing to diffraction-limited with 100-m. All images 0.6 × 0.6 arc seconds²

Simulations show that V=38 point sources can be detected in 10 hours assuming diffraction-limited quality, more than a 1000-fold improvement over the VLT. OWL is very complementary to NGST. NGST has unmatched capability in the thermal IR, while OWL is better for imagery at $\lambda < 2.5\mu m$ and spectrography ($R \geq 5,000$) at $\lambda < 5\mu m$. Sensitivity-wise, OWL could not compete in the thermal IR, but has a much higher spatial resolution. OWL would also have a synergetic role with ALMA (e.g. in finding and/or studying proto-planets) and with VLBI (the radio astronomers have been waiting for us optical/IR people to catch up in spatial resolution for decades!)

1.5 Interferometry. Is Interferometry an alternative to a filled aperture concept? In fact, it has a clearly separate scientific niche – for similar baselines, its field of view (few arcseconds) and magnitude limits are not competitive with a filled aperture telescope. On the other hand, kilometer-size baselines (hundreds of km in space) might well be reached in the future. Looking for the details of bright objects at the micro-arcsecond level, discovering earth-like planets, studying the surface of more distant stars are privileged domains for Interferometry.

2 Items for the Science Case

The science case for future extremely large telescopes is not fully developed yet, even if virtually every branch of astronomy should be deeply affected by the availability of a 100-m telescope with the characteristics outlined above. This is one main component of our ongoing effort, and is crucial to derive the final set of requirements for OWL, e.g. its wavelength coverage and degree of adaptive optics correction. In the following we discuss some areas where OWL could give unprecedented contributions.

Confusion about confusion. There is a widespread concern that ELTs may hit the confusion limit, thereby voiding their very *raison d'être*. Much of it comes from past observations at poor angular resolution (e.g. X-ray data or deep optical images in $2''$ seeing of the '80s). Recent results with better resolution lead to resolving it in individual objects (e.g. the X-ray background, now mostly resolved, or the HDF images showing 20 times more empty space than objects). Ultimately, some confusion level will be reached, but the 3-dimensional nature of astronomical objects (position and velocity) virtually ensures it will not be a limiting factor with OWL.

Star formation history of the Universe. The history of stellar formation in the Universe from the first stars to present day is one of the 'hot topics' in astrophysics. "Measurements" of star formation rates are obtained at a variety of look-back times, comparing an integrated light flux with predictions made by evolution models. With OWL these "fuzzy" objects would be resolved in their stellar components (Figure 2). One could see O stars out to a redshift $z \sim 2$, HII regions at $z \sim 3$, SNe out to $z \sim 10$. Determining star formation rates in individual galaxies would now be directly obtained from the study of individual components, much as presently done for nearby galaxies.

Measure of H. Cepheids could be measured with OWL out to a distance modulus $(m - M) \sim 43$ (i.e. $z \sim 0.8$). This would allow the direct measurement of H (not Ho) and its dependence on redshift, allowing a true measure of the scales of the Universe.

Supernovae at $z \sim 10$. Extrapolating from even intrinsically faint nearby supernovae like SN 1987A, OWL appears able to detect all SN up to $z \sim 10$, i.e. probably any supernova that ever exploded (!). Model calculations give an expected rate of 2 supernovae per week in massive elliptical galaxies, with a lifetime of a few days in the redshifted UV. This means that any deep exposure in a field ≤ 1 arcmin^2 will contain several new supernovae, giving a distance-unbiased full history of massive star formation.

Other high redshift Universe studies. A telescope with OWL's resolution and sensitivity would find some of its most important applications in the study of the farthest and faintest objects in the Universe, with in particular studies of the proto-galactic building blocks and their merging into higher hierarchical structures. The possibility of probing even larger redshifts with Gamma Ray Bursts (if they exist at earlier epochs) is very exciting, as they are intrinsically orders of magnitude brighter than even SNe.

High frequency phenomena. Rapid variability is an area where the improvements brought by larger collecting areas is very large, by $\sim D^4$. ELTs will open a window on the study of quantum phenomena, till now only observed in the laboratory.

Nearby Universe. There is again a myriad of possible contributions. It would be possible to observe brown dwarfs in the Magellanic Clouds, white dwarfs in the Andromeda galaxy and solar-like stars in galaxies in the Virgo cluster, enabling detailed studies of stellar populations in a large variety of galaxies. The environment of several AGNs would be resolved, and the inner parts nearest to the central black hole understood. In our own galaxy, we could study star formation regions like Orion at sub-AU scales, detect circumstellar disks and possibly proto-planets, and image the surface of hundreds of stars, promoting them from points to objects.

Extra-solar planets. Finally, a critical contribution will be in the subject of extra-solar planets. Not so much in their discovery (interferometry should be quite successful in this), but for spectroscopic studies. Measuring their chemical composition, looking for possible biospheres will be one of the great goals of the next generation of ELTs. Figure 2 shows a simulation of an observation of the Solar System at 10 parsecs (including the effect of micro-roughness and dust diffusion on the mirrors) where Jupiter and Saturn would be readily detected. Several exposures, preferably with sophisticated coronographic techniques, would be necessary to detect earth-like planets.

Fig. 2. Simulation of the Solar System at 10 parsecs. Jupiter can be seen on the right. Saturn would also be detected, about 10 cm on the right of this page

Observational strategies. The sheer size of the OWL facility makes it unlikely that its observing scenario would be similar to that of the current generation of telescopes. The current (mild) trend towards Large Programs (where the need for deep – i.e. long – exposures is combined with the statistical requirement of a large number of measurements) will likely evolve towards a "Big Project" approach, as is currently the case in particle physics. Such a Project may develop the "best" instruments for its observations, and when completed a new Project with possibly another instrument set would take over. What we envision are "seasons" in which OWL will e.g. image the surface of all accessible stars, or study 10^5 SNe, or follow the disruption of a star by a supermassive black hole. In other words, a series of self-contained programs which tackle (and hopefully solve!) well defined problems, one at a time. A posteriori discoveries by mining the huge volume of data generated by these specialized instruments would also become the norm.

3 Telescope Design

New generations of telescopes built over the last century, from the Hooker to VLT and Keck have generally relied on extensive R&D programs, in particular in the area of optics fabrication. Optical configurations basically did not change: with few exceptions, telescopes were designed on the basis of a Ritchey-Chrétien solution. The new emphasis has been put on solving optical scalability problems (new materials, active optics, segmentation), reducing cost (e.g. enclosure costs: alt-az-mount, faster primary mirror hence shorter telescope length), relaxing tolerances and improving performance (active optics).

NTT, VLT, HET and, to a lower extent, Keck, embody a certain rethinking of system aspects of telescope design and construction. To a great extent, these projects have relied on industry for the design and fabrication of crucial components, while developing expertise and solutions at system level. OWL could be seen as a step further in the direction of an industrial approach: its design shifts the balance in the direction of proven subsystem technology, hence low industrial risk, while allowing for some (limited) system innovation, e.g. the combination of active and segmented optics, and modular mechanical design. This approach has already led to a positive response of industry to the OWL concept.

A crucial objective pursued in the design of OWL is predictability, i.e. reliable optical, mechanical and control solutions. The experience derived from VLT (performance and reliability of active optics), Keck (virtually unlimited optical scalability), and HET (spherical primary solution) ensures that this objective can be met in the field of telescope optics, as the associated processes and design solutions are fully demonstrated.

The design, fabrication and integration of mechanical structures are inherently predictable, as accurate modeling is possible. The same may be said of control systems (the number of degrees of freedom to control in a telescope like OWL –a few thousands, at modest frequencies– may appear unusual for an astronomical telescope but does not require any technology breakthrough).

Under these conditions, it appears that the scalability of telescope diameter is no longer set by optical fabrication processes, but most likely by structural characteristics and cost. Our estimate is that the technical limit for a fully steerable telescope is ∼130-150-m, possibly more if high performance (and prohibitively expensive) materials are used for the telescope structure.

As of today, the evident exception to suitable predictability is adaptive optics. The fact that the principle of multi-conjugate adaptive optics has been verified on the sky, together with the pressure implied by its promises, and potential spin-off from consumer applications (MEMs), leave room for (cautious) optimism.

3.1 Design Evolution

The evolution of the telescope design has been marked by a few key trade-offs and subsequent decisions: (i) Segmented primary and secondary mirrors, segment dimension ~ 2-m; (ii) Optical design, spherical primary mirror solution; (iii) Non-adaptive main telescope optics; (iv) Implementation of active optics and field stabilization; (v) Alt-az mount.

Segmentation of the primary mirror is evidently required; large (4 to 8-m) segments have been excluded mainly for cost and risk reasons (8-m class segments would imply a primary mirror cost of ~ 0.8-1.0 billion Euros, support not included). The maximum size has been set to ~2.3-m for cost-effective transport (standard containers). The exact dimensions are still to be defined and will mostly be determined by performance, reliability and cost criteria. Preliminary analyses indicate that the optimal dimension is probably 2-m.

Several optical designs have been explored, from Ritchey-Chrétien solutions to siderostat with relatively slow primary mirror. It has been found that cost, reliability, fabrication and telescopes functionality considerations point towards a spherical primary mirror solution. In terms of fabrication, all-identical spherical segments are ideally suited for mass-production and suitable fabrication processes recommended by potential suppliers are fully demonstrated. In a modified version, these processes could also be applied to aspheric segments, however with a substantial cost overshoot (more complex polishing, lower predictability, tighter material requirements). It can be shown that an aspherical solution also implies faster primary mirror for equivalent structure height, thereby exacerbating telescope alignment and mirror fabrication issues. The secondary mirror becomes a critical issue as well. If convex (e.g. Ritchey-Chrtien solution), it must be small (~2-3-m) in order to be feasible (optical testing), which again implies tight alignment tolerances (at a location where they are the least achievable). A Gregorian solution solves the secondary mirror feasibility problem, however at the cost of a longer telescope structure.

The current baseline design is based on a spherical primary mirror solution, with flat secondary mirror and aspheric, active corrector. The flat secondary mirror has also major advantage in terms of decenters (cm rather than μm), which are evidently crucial with a structure the size of OWL's.

Although in principle very attractive, the option of incorporating adaptive mirrors into the main telescope optics has been ruled out, mainly for reliability and cost reasons. The best figure announced so far for large adaptive secondary mirrors is around 1000 Euros per degree of freedom (mirror *not* included); with 50,000 to 500,000 degrees of freedom per adaptive mirror, as required with OWL, the cost implication becomes unbearable. Cost-effective solutions imply relatively small mirrors and are inherently more reliable (small moving masses).

In order to allow seeing-limited operation and to avoid over-constraining adaptive systems, the telescope itself incorporates several wavefront control functions, in particular segments phasing, active optics and field stabilization.

Several mount options have been assessed, from a classical alt-az solution to de-coupled primary and secondary mirror structures. Cost and performance considerations point clearly towards the alt-az solution (figure 3). Although it is tempting to "borrow" concepts from radio-telescope designs, it should be born in mind that radio telescopes have extremely fast reflectors, which allow for a much shorter structure. The current baseline design is modular i.e. the structure is made of (nearly) all-identical, pre-assembled modules. This crucial feature allows for major cost savings. It should also be noted that, in proportion to size, the OWL structure is one order of magnitude lighter than traditional telescope ones.

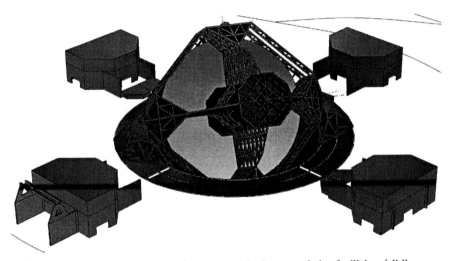

Fig. 3. Telescope pointing at 60° from zenith, layout of the facilities (sliding enclosure not shown)

There is no provision for a co-rotating enclosure, the advantage of which is anyway dubious in view of the enormous opening such enclosure would have. Protection against adverse environmental conditions and excessive daytime heating would be ensured by a sliding hangar, whose dimensions may be unusual in astronomy but actually comparable to or lower than those of large movable enclosures built for a variety of applications. Mirror covers sliding into the structure would provide segments cleaning and handling facilities, and local air conditioning if required.

Relevant site aspects are more complex than with previous telescope generations, mainly because of multi-conjugate adaptive optics (MCAO), telescope size, and the higher impact of seismic activity on cost and safety. MCAO implies, in particular, that the function of merit of the atmosphere cannot be described by a single parameter (seeing). Better understanding of site quality in relation to climatology is also essential. A positive development is the availability of databases providing suitable worldwide coverage. Indeed, the first step in a site selection process will be to screen those databases with respect to relevant parameters.

3.2 Cost and Schedule

Cost estimates indicate that the required capital investment could be on the order of 1 billion Euros, including contingency. These estimates are, however, extremely preliminary; industrial studies are required for consolidation. Compared to "classical" telescope cost factors, substantial cost reduction occurs with the main optics (fabrication processes adapted to mass-production), the telescope structure (very low mass in proportion to dimensions, mass-produced modules), and the enclosure (reduced functionality, no air conditioning).

Preliminary schedule estimates indicate that technical first light could plausibly occur within 8-9 years after project funding. Allowing for 2.5 years integration and verification of the IR adaptive module(s) and 3.5 years for integration and verification of the visible adaptive module(s), the telescope could already deliver science data in the IR and in the visible within 10.5-11.5 and 11.5-12.5 years after project go-ahead, with unmatched resolution and collecting power. Full completion would occur ~15 years after project start. The critical path to first light is set by the structure and the enclosure.

4 Conclusions

From the technical point of view, no obvious showstoppers to build a 100-m telescope have been identified so far. The price tag of many ELTs remains below the cost of a medium space mission, so we could call it "reasonable". The timeline for construction is around 10 years. Industry is indicating that there is an interest in building one, and that they agree about its feasibility. The science case is exciting and stunning, and there is an unmatched potential for new discoveries. Let's do it!

Acknowledgements

OWL is the brainchild of the OWL people. Special thanks to E. Brunetto, B. Delabre, N. Hubin, F. Koch, J. Spyromilio, M. Quattri for the many discussions and their work.

The Next Generation Space Telescope

Steven V.W. Beckwith

Space Telescope Science Institute, 3700 San Martin Drive,
Baltimore, MD 21218, USA

Abstract. The Next Generation Space Telescope is designed to study the most distant objects in the universe, the first stars and galaxies that were born after the Big Bang. As a logical successor to the Hubble Space Telescope, it will stress deep imaging and multi-object spectroscopy in the near and thermal infrared portions of the spectrum. Results from the Hubble Deep Field show that galaxy formation was vigorous at redshifts between 1 and 5 and that galaxies looked different at that epoch. NGST is designed to observe the details of galaxy genesis and buildup out to redshifts of 10. It will be sensitive enough to see back to the time before galaxies were born.

1 Introduction

For the last four hundred years, telescopes have been our window on the universe. The first telescopes used by Galileo almost instantaneously changed humandkind's view of its place in the universe. His discovery of the moons of Jupiter demonstrated that the Earth was not the sole center of orbiting bodies; he saw mountains on the Moon, letting him make a reasonable estimate of the Moon's distance; he saw that the Milky Way breaks up into many stars when viewed at high magnification, immediately increasing the number of known stars by an enormous factor. Most important was his observation that Venus displayed phases like the Moon. An important prediction of the Copernican theory was that Venus should display phases, so Galileo's discovery more than any other piece of evidence supported a Sun-centered universe (e.g. Drake 1984).

Of these discoveries, only the phases of Venus were predicted beforehand as a test of a scientific theory. The others came naturally because of the advance that the telescope brought to observing the sky. Galileo's telescopes improved the sensitivity of the unaided human eye by almost two orders of magnitude. Perhaps more important was the improvement in angular resolution; those early telescopes increased the resolving power of the eye by a factor of about 30. These improvements brought new phenomena into view that were not imagined prior to their discovery - or at least they were among so many possible phenomena that could be imagined that they played no role in predictive theories.

The Hubble Space Telescope brought a similar two orders of magnitude gain in our sensitivity to distant objects. The finest expression of this sensitivity is the observation of the Hubble Deep Field (HDF), a 10 day exposure of the sky that is our deepest glimpse into the heavens (Williams et al. 1996). The HDF revealed a large population of galaxies at redshifts above 1 and provided us with a sample of the early universe as it looked soon after the first stars and galaxies came into existence. It is our best example of what we can expect to see when we improve our telescopes by another two orders of magnitude.

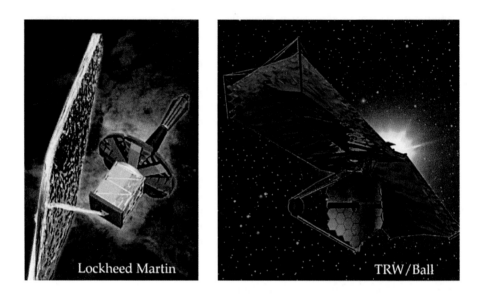

Fig. 1. Two possible designs for the NGST are the one by Lockheed/Martin, shown on the left, and the one by TRW/Ball, shown on the right.

The Next Generation Space Telescope (NGST) will be the successor to Hubble. It is currently planned to have an aperture diameter of 8 m compared to Hubble's 2.4 m. It will be cooled to 40 K to reduce its thermal emission, and it will be in orbit around the Earth-Sun system at the Lagrangian point, L2, making it a superb telescope for wavelengths in the 1—30 μm range (see http://www.ngst.stsci.edu and http://ngst.gsfc.nasa.gov). NGST is designed to study the early universe in detail, to look back to a time when the first stars and galaxies were born after the Big Bang. It will do that with exquisite precision. Because it will increase our observational capability by a similar factor to what Galileo's telescope and the Hubble Space Telescope did in their times, we can expect new discoveries to be made. NGST will usher in a new era of discovery as we explore the outer reaches of the universe.

2 The Hubble Deep Fields

The Hubble Deep Fields, one in the north (Williams et al. 1996) and one in the south (Williams et al. 2001), uncovered a population of galaxies at redshifts between 1 and 5. The HDF images have enough resolution to discern their size, shape, and some structure for comparison to galaxies nearby. For the currently popular flat universe with 30% of the energy density in matter and 70% in a cosmological constant (vacuum pressure), the universe was between 45% (z=1) and 12% (z=5) of its present age. These images reveal differences between early galaxies and the ones in the local universe apparent by inspection. Ferguson, Dickinson, and Williams (2001, hereafter FDW01) review the advances made through these images, a few of which are summarized here. A complete list of original references to conclusions drawn from this work are in the FDW01 review. For the purposes of this paper, "high redshift" will refer to redshifts above 1, and "early universe" refers to the universe when it was less than about 40% of its present age.

In the early universe, there were many irregular galaxies and relatively few that were "grand design" spirals or ellipticals. A plot of the fraction of spirals, ellipticals and irregulars shows a slow growth of the first two types and a rapid growth of the irregulars as a function of magnitude going to faint magnitudes (Abraham et al. 1996). A check of the redshifts shows that fainter does, indeed, mean higher redshift, so the increase of irregulars is due to a real change in the shapes of galaxies in the early universe, not the discovery of a local population of faint, low-surface brightness galaxies. These results are not without some controversy, because the images of the highest redshift objects are at short (ultraviolet) rest-frame wavelengths where even local galaxies often appear less regular than they do at longer wavelengths. Nevertheless, this bandpass effect probably only accounts for some of the irregularity and does not dominate the trend (see below).

The increasing number of irregular shapes is consistent with hierarchical clustering (Baugh et al. 1996) as the mechanism by which galaxies were assembled. In hierarchical clustering, the earliest collections of stars were irregular clumps brought together by gravity. These clumps merged with others to create large galaxies which subsequently evolved to smooth out the irregularities and elicit the ordered motion associated with relaxed ellipticals or grand design spirals driven by ordered wave motion. Looking back to when the universe was only about 25% of its present age, we see that irregular shapes dominate the population of galaxies.

Early galaxies were apparently smaller than they are today (Simard et al. 1999; Lowenthal et al. 1997; Roche et al. 1998). High redshift galaxies in the HDFs appear to be physically smaller than local galaxies, after allowing for our inability to discern low surface brightnesses in high redshift objects. This trend emerges from a visual inspection of the images and from simulations of local galaxies as they would appear at high redshift compared to the HDF galaxies at redshifts greater than about 1. If galaxies were built up by hierarchical clustering, they should have been smaller in the early universe. The small apparent sizes are further evidence that the assembly of galaxies took place at redshifts greater than about 2.

The shapes of high redshift galaxies are also different from their low-redshift counterparts, but here one must worry about distortions caused by comparing galaxies at different wavelengths (FDW01). Optical images of the HDF correspond to rest-frame ultraviolet images for galaxies at redshifts of 2 or more. It is well known that the ultraviolet appearance of local galaxies differs from the optical appearance, owing to the combined effects of variable extinction – which is stronger in the ultraviolet – and the dominance of hot, blue stars that are seen only in regions of star formation. The ultraviolet appearance of a local grand design spiral may look nothing like a spiral but instead consist of a series of clumps and knots coincident with those star forming regions with patchy extinction.

Infrared images of the HDF with NICMOS sample the galaxies at the rest frame optical and far red wavelengths. In most cases, the irregular, knotty appearance of the high redshift galaxies does not change with wavelength (Teplitz et al. 1998; Thompson et al. 1999; Dickinson et al. 2000). The similar appearance between rest frame UV and optical suggests that the irregular shapes result from irregular distributions of matter, not simply a change in the way the light is emitted. These shapes are also what one expects in a hierarchical clustering universe in which the pieces at the beginning – our high redshift galaxies – are smaller and less regular than the final products.

It is fairly well established that the rates of star formation were greater at redshifts of 1 to 3 than they are today. Studies of the HDF show that star formation in the early universe was at least ten times what it is today (Madau et al. 1996; Madau 1997; Dickinson 1998; Steidel et al. 1999). There is still controversy about whether these high rates of star formation started as early as redshift 5. Initial estimates of the star formation rate as a function of redshift or lookback time indicated a peak in the rate at a redshift between 1 and 3 with a decreased rate at higher redshifts. The decrease at higher redshifts would be consistent with most galaxy formation happening at redshifts of a few from a slow buildup from earlier times.

This peak has been questioned by a number of studies that attempt to correct the observed light for extinction, which tends to raise the estimates at high redshifts relative to those at low redshift. The high redshift light comes from the rest frame ultraviolet, and the corrections for even modest extinction become large. The production of metals from this early star formation is broadly consistent with observations in the intergalactic medium through quasar absorption lines for either case (Calzetti & Heckman 1999; Pei, Fall, & Hauser 1999). Although we cannot be certain at present whether the star formation rates really are maximal at $1 \leq z \leq 3$ or flat out to the highest redshifts observed, it is certain that we are witnessing an important epoch in galaxy formation in the HDF images alone.

How do the galaxies relate to the creation of giant black holes in active galactic nuclei? The density of AGN as a function of redshift rises from the present epoch to a broad peak at a redshift around 3, then falls toward higher redshifts. There is a possibility that the falloff at high redshift will disappear when we are able to discover obscured AGN through x-ray surveys; indeed, there are hints of this possibility at this conference from Hasinger's presentation (Hasinger 2001, this conference). Whether the density of AGN as a function of redshift is peaked or flat, it is similar enough to the behavior of the star formation rate to suggest that the buildup of black holes and the buildup of galaxies occured nearly simultaneously, even though we are comparing a rate to a density. The understanding of how AGN and galaxies are linked is a key question for future research.

Finally, the total cosmic energy density in the far infrared should be an integral of the energy released during galaxy buildup in the early universe. The measurements of the infrared background radiation from COBE were a monumental achievement (Hauser et al. 1998; Hauser & Dwek 2001). They suggest that we are still missing some energy, if the HDFs are typical of the rate of energy release inferred from the star formation rates. However, the uncertainties are large enough to allow consistency between the energy released from stars that we can see (or extrapolate with conservative assumptions) and the infrared background as measured by COBE.

An interesting surprise in the HDF north was the discovery of an object seen only in the near infrared starting at $1.6\,\mu$m. This "J dropout" object is unusual in that it appears slightly extended and very red. It may be a young galaxy at a redshift near 12, or it may be an evolved elliptical galaxy at a redshift around 3 but reddened by dust (Dickinson et al. 2000). It portends other discoveries to be made in future deep surveys in the infrared. Like much cutting edge science, it is too faint to be studied properly with this generation of telescopes, and it will have to await the next.

3 Future Surveys

By extrapolating the source counts from the HDFs, we can estimate the impact of future surveys for high redshift objects. Figure 2 shows number counts as a function of flux density limits for galaxies and AGN at different redshifts. The HDF has already reached the practical limit for Hubble with its current instrumentation. With the addition of the Advanced Camera for Surveys (ACS) and the Wide Field Camera 3 (WFC3), we can increase the number counts of high redshift objects by about a factor of five. An ACS deep field should yield several hundred galaxies with redshifts between 5 and 10 and perhaps 10 galaxies at redshifts above 10. The number of AGN would be much lower. Chandra should discover many AGN at redshifts greater than 5, but without followup observations to establish the redshifts and Hubble images to understand the source morphology, it will be difficult to get a good idea of the population's characteristics.

NGST is designed to open up the high redshift universe to easy observation. An NGST deep field should discover more than 10^4 galaxies and 1000 AGN in the range $5 \leq z \leq 10$. It will see more than 1000 galaxies at redshifts beyond 10, assuming the extrapolations can be trusted. Such large samples would revolutionize the study of galaxy formation.

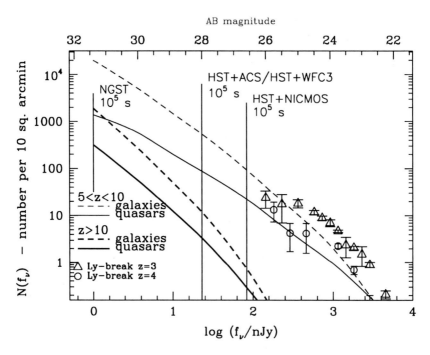

Fig. 2. This plot is from Harry Ferguson, showing the number counts of Lyman break galaxies at $z \sim 3$ and $z \sim 4$ from Steidel et al. 1999, (triangles and circles, respectively) together with predictions of number counts at higher redshift from Haiman and Loeb (1998). The limits for the HDFs, a deep survey with ACS and WFC3, and an NGST deep survey are shown.

A more important point is that galaxies at redshifts beyond five move out of the visible range available to optical CCD detectors. At a redshift of five, for example, essentially all the rest frame energy is at wavelengths beyond about $1\,\mu$m. The WFC3 will be the only instrument with enough sensitivity to see these galaxies, and it will observe the rest frame ultraviolet. To understand the structure, metallicity, star formation rate and history of these galaxies, we will need an infrared telescope like NGST.

4 The Next Generation Space Telescope

NGST will be designed primarily to study the high redshift universe. With an aperture of 8 m, reduced thermal emission from cooling and its L2 orbit, NGST will have two to three orders of magnitude more sensitivity than comparable sized telescopes on Earth. It will offer a large improvement over Hubble, too, although there will be only modest overlap in the wavelength ranges covered by Hubble (0.1—2 μm) and NGST (0.6—27 μm). It will complement Hubble's ability to study galaxies out to redshifts of 5 with an ability to study galaxies to redshifts beyond 10. Furthermore, Hubble is already limited by wavelength coverage and sensitivity in its observations of galaxies beyond redshifts of about 2, where most of the action in galaxy birth takes place. NGST will offer a combination of excellent sensitivity and superb resolution to study the universe when it was less than half its present age, at redshifts beyond about 1.

There are currently three instruments planned for NGST. The first is a near-infrared imager. This camera will operate between about 0.6 and 5 μm, the wavelength range accessible by InSb detectors. It is planned with a field of view of four minutes of arc on a side, meaning an $8k \times 8k$ pixel array to provide a Nyquist sampled image over the entire field. In addition to imaging, it should be equipped with a grism allowing low resolution spectroscopy with a resolving power ($\frac{\lambda}{\Delta\lambda}$) of about 100. This imager will be the workhorse instrument for NGST, just as WFPC2 has been the dominant instrument on Hubble.

The second instrument will be a near infrared multi-object spectrometer. This spectrometer is planned to observe more than 100 objects simultaneously over a three minute of arc field of view with a resolving power of 1000 or more. The wavelength range will be 1—5 μm, similar to the near infrared imager.

The third instrument will be thermal infrared camera/spectrometer. Covering the wavelength range 5—27 μm, it should be Nyquist sampled at 10 μm over a two minute of arc field of view. The wavelength range is set by the current sensitivity of arsenic-doped silicon (Si:As) detectors, although it could be extended if some of the newer doping materials (e.g. Si:P) are mature enough for selection within the next few years. Since this will be the only thermal infrared instrument, a goal is to have spectroscopic capability, too, with resolving powers of order 1500. It is not clear if multi-object capability is possible for an ambitious instrument of this sort. If focal plane devices such as micro-electronic mirrors or slits now under development reach maturity, it may be possible to incorporate multi-object capability into the instrument without a large impact on cost and complexity.

These instruments have great potential to address a variety of important science questions in astronomy in addition to the high redshift universe. The near infrared imager will also study dark matter, distant supernova, young stars, Kuiper Belt Objects, and the populations of stars in our own and other local galaxies. If it has a coronographic capability, it will observe faint brown dwarf stars and giant planets in the solar neighborhood. The multi-object spectrometer will study galaxy formation in detail, including abundances, star formation rates and velocity distributions, as well as AGN. It will be ideal for studying young stellar clusters in our galaxies to understand the initial mass function and will be unique for spectra of the atmospheres of brown dwarf stars and free-floating giant planets. The thermal infrared instrument is well suited to the study of evolved stars at high redshift, obscured star formation regions at redshifts around 5, AGN to the same high redshift, PAH features in galaxies out to redshifts of 5, H_α to $z \sim 15$, the initial mass function of cool stars, the disks around protostars in the Galaxy, the sizes of Kuiper Belt Objects, and distant comets, to name but a few. This suite of instruments will be a powerful enhancement to humankind's ability to study the universe.

It is worthwhile describing four key science programs that NGST will carry out uniquely. The first is the NGST Deep Field (NDF). An NDF would cover a $4' \times 4'$ field, four times the area of an HDF, to an AB magnitude of 34, more than 100 times deeper than Hubble can do. Within this field, we expect to see 5000 galaxies to an AB magnitude of 28, and more than 10^5 galaxies at the faintest magnitudes. For each of these galaxies, we will observe the morphology, photometry, and photometric redshifts. It will be possible to see Ly_α emission-line galaxies to a redshift of 40; the 4000 Å break could be detected out to redshifts of 10. It will be easy to see starburst galaxies that create stars at $1\,M_\odot\,yr^{-1}$ for only 1 Myr out to redshifts of 20. It will be possible to observe galaxies forming stars at $0.1\,M_\odot\,yr^{-1}$ for 10^9 yr out to redshifts beyond 10. A deep field should revolutionize the study of high redshift galaxies and uncover the first generation of stars in the early evolution of the universe.

NGST will be able to measure precisely the epoch of reionization of the universe. We know the universe is highly ionized out to redshifts of at least 5. But it had to be neutral at very high redshifts owing to an absence of ionizing stars or black holes. At present, the redshift at which the universe was reionized is a mystery. NGST will observe the redshift of reionization directly, and it will be unique for this task, if reionization occurs at a wavelength that the Earth's atmosphere does not transmit, at $z \sim 8$ or 11, say. The basic technique will be to discover objects at redshifts beyond 10, then look for the absorption edge at the place where hydrogen was neutral in the intergalactic medium. Figure 3 shows a simulated spectrum of a high redshift object with the absorption edge clearly marked.

A third key project will be to discover and study high redshift supernovae to use as standard candles for measuring the expansion rate of the universe. One of the most interesting discoveries of the last decade was that the universe appears to be accelerating its expansion, presumably because the cosmological constant is non-zero (but not enormous, either; Schmidt et al. 1998; Perlmutter et al. 1999). The discovery relies on using distant supernovae as standard candles to measure the distance as a function of redshift. A

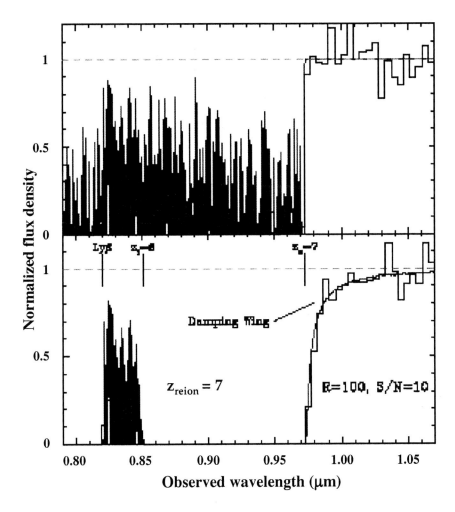

Fig. 3. This figure is a simulated spectrum of a high redshift object if reionization of the universe occurs at a redshift of 7.

problem is that with the best ground and space based telescopes, it is difficult to discover supernovae at redshifts beyond about 1, and that is just where the curves distinguishing different cosmological parameters begin to diverge. NGST will be able to detect supernovae of the appropriate standard candle type out to redshifts of 5 or greater. At these redshifts, it will be possible to see if the expansion rate follows one of the regular cosmological curves or if there is some other effect that may be giving us false indications of the supernova brightness.

Figure 4 plots the luminosity distance against redshift for the known supernovae (inset) and shows the extent to which NGST can discriminate

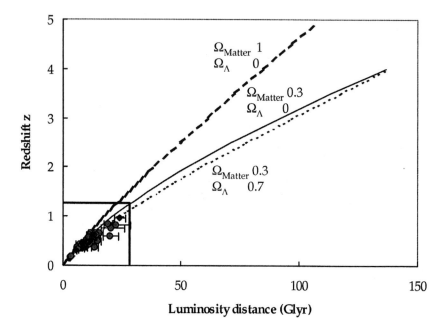

Fig. 4. The figure shows the luminosity distance versus redshift for supernovae in different cosmologies. The box in the lower left hand corner shows the range of redshifts available to HST and ground-based telescopes along with data from the two groups assembling high redshift supernovae (Schmidt et al. 1998; Perlmutter et al. 1999). The curves show the expected behaviour of the data for different cosmologies. NGST can study supernovae out to redshifts beyond 5. Note that NGST can easily distinguish between cosmologies in a flat universe ($\Omega_{matter} + \Omega_\Lambda = 1$), although it is less sensitive to changes in any one cosmological parameter, such as Ω_Λ.

among different cosmological curves. This method is sensitive to the amount of matter in the universe, Ω_m, and less sensitive to the cosmological constant itself. The figure shows two curves that diverge around $z \sim 1$ and come together again at high redshift for different values of Ω_Λ. The data indicate divergence at low redshift, and it would be essential to follow one of these curves to demonstrate clearly that the trend is cosmological and not intrinsic to the supernovae brightness.

Finally, when the density of objects seen at high redshift is very large, gravitational lensing by foreground matter becomes important for every image. Figure 5 is a simulation of the shapes of galaxies in a long exposure such as an NGST Deep Field. A glance at this figure shows the obvious distortion introduced by the foreground matter distribution. Since almost all of this matter is dark, the lensing is one of the few ways to measure the dark matter distribution directly. By modeling the dark matter distribution to deconvolve

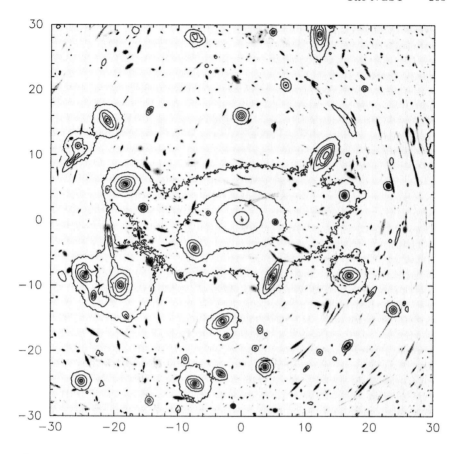

Fig. 5. This is a simulated image from the central part of the cluster Abell 2218 taken from the NGST Design Reference Mission, courtesy of Peter Schneider. Details of the simulation can be found at http://www.ngst.stsci.edu/drm/programs.html. The strong lensing of the background galaxies is apparent in the image, where the contours show the dark matter distribution.

these images and produce distortion-free pictures, it should be possible to derive the dark matter distribution reasonably well in a statistical sense and, therefore, directly measure the matter content of the universe. NGST will be unique for these measurements, since they rely on having large numbers of high redshift objects for studies of the lensing on fine scales.

Acknowledgements

Many of the arguments in this contributions came from conversations with and presentations from Harry Ferguson, and he contributed Figure 2; I am pleased to thank him for his help. I also thank Peter Stockman and John Mather for their insight that I gleaned from various presentations on the Next Generation Space Telescope, and Adam Riess who provided advice and data for Figure 4.

References

1. Abraham, R. G., Ranvir, N. R., Santiago, B. X., Ellis, R. S., Glazebrook, K., & van den Bergh, S. (1996) MNRAS **279**, L47
2. Baugh, C. M., Cole, S., & Frenk, C. S. (1996), MNRAS, **282**, L27
3. Calzetti, D. & Heckman, T. M. (1999), Ap. J. **519**, 27
4. Dickinson, M. (1998), in *The Hubble Deep Field*, ed. M. Livio, S. M. Fall, & P. Madau, (Cambridge: Cambridge University Press), p. 219
5. Dickinson, M., Hanley, C., Elston, R., Eisenhardt, P. R., Sanford, S. A., et al. (2000), Ap. J. **531**, 624.
6. Dickinson, M. et al. (2000), astro-ph/0004028
7. Drake, S. (1984), Jour. Hist. Astron. **xv**, 198
8. Ferguson, H. C., Dickinson, M., & Williams, R. E. (2001), Ann. Rev. Astron. Astroph. **38**, in press; astro-ph/0004319
9. Haiman, Z. & Loeb, A. (1998), Ap. J. **503**, 505
10. Hauser, M., Arendt, R. G., Kelsall, T., Dwek, E., Odegard, N., et al. (1998), Ap. J. **508**, 25
11. Hauser, M. & Dwek, E. (2001), Ann. Rev. Astron. Astroph., in press
12. Lowenthal, J. D., Koo, D. C., Guzman, R., Gallego, J., Philips, A. C., et al. (1997), Ap. J. **481** 673
13. Madau, P. (1997), in *Star Formation Near and Far*, ed. S. S. Holt & L. G. Mundy, (Woodbury, NY: AIP Press), p. 481
14. Madau, P., Ferguson, H. C., Dickinson, M., Giavalisco, M., Steidel, C. C., & Fruchter, A. S. (1996), MNRAS **283**, 1388
15. Pei, Y. C., Fall, S. M., & Hauser, M. (1999), Ap. J. **522**, 604
16. Perlmutter, S., Aldering, G., Goldhaber, G., Knop, R. A., Nugent, P., et al. (1999), Ap. J. **517**, 565
17. Roche, N., Ratnatunga, K., Griffiths, R. E., Im, M., & Naim, A. (1998) MNRAS **293**, 157
18. Schmidt, B. P., Suntzeff, N. B., Phillips, N. N., Schommer, R. A., Clocchiatti, A., et al. (1998), Ap. J. Suppl. **507**, 46
19. Simard, L., Koo, D. C., Faber, S. M., Sarajedini, V. L., Vogt, N. P., et al. (1999), Ap. J., **519**, 563
20. Steidel, C. C., Adelberger, K. L., Giavalisco, M., Dickinson, M., & Pettini, M. (1999), Ap. J. **519**, 1
21. Teplitz, H., Gardner, J., Malmuth, E., & Heap, S. (1998), Ap. J. Lett. **507**, L17
22. Thompson, R. I., Storrie-Lombardi, L. J., Weymann, R. J., Rieke, M. J., Schneider, G., et al. (1999), Astron. J. **117** 17
23. Williams, R. E., Baum, S. A., Bergeron, L. E., Bernstein, N., Blacker, B. S., et all (2001), Astron. J., in press
24. Williams, R. E., Blacker, B., Dickinson, M., Dixon, W. V. D., Ferguson, H. C., et al. (1996), Astron. J. **112**, 1335

The Essential Role of Space Astronomy

R.M. Bonnet

Director of the Scientific Programme
European Space Agency

1 Introduction

It is a great honour for me to be here, addressing one of the most talented and renowned astronomers of this century on the occasion of his 70th birthday. Referring to Lo Woltjer, age has no meaning. It is another honour to be the last to speak among so famous people and to close what has been a fascinating symposium. My task is not made easier by the title of this paper which sounds a little strange to celebrate the former Director General of one of the most prestigious Earth-bound astronomical observatories. However, the deep involvement and pioneering contribution of Lo Woltjer in the interpretation of the X-ray background, renders my talk a little easier.

In the course of the meeting a large number of references were made to the role space techniques have played in astronomy and it is not easy at this point to add any new or original contribution. I therefore apologise if it seems that my words appear to be a repetition of what has been said already. Hopefully, there are still a few things to be mentioned which can be exploited now to emphasize even more both the role of space astronomy and the essential role of Lo Woltjer in the field. But before discussing that, we should perhaps come first to a more basic question.

2 What Is Space Astronomy?

For a very long time, astronomy was the observation of the "visible" sky and obviously of the sky in the visible part of the electromagnetic spectrum. Nevertheless, from the time when we used only our eyes and the time of the first telescopes of Galileo and Newton, astronomy has become more and more the science of the "invisible" sky. Invisible in the sense that the objects are too faint to be observed directly, or because our telescopes are too small. But invisible also because their light does not reach the ground as it is absorbed by the Earth's atmosphere. That part of the "invisible" is the realm of space astronomy since we need to go into space to detect those photons which never reach the ground. Once in space, there are no limitations in principle to our observing capability and we can even dream of exploring in situ some of the closest of the celestial sources which dotted the skies of our childhood. Space techniques offer the unprecedented capability of observing all objects in the Universe that emit, reflect, and scatter light, from the Big Bang itself through galaxies, quasars, stars, planets but also dust and a growing number of objects so massive and so small that they can only be black holes.

Often shown in the course of this meeting has been the energy density spectrum of the Universe over the whole electromagnetic spectrum from radio to gamma rays and reproduced here in Fig. 1 (Hasinger 2000). This truly remarkable figure clearly evidences that in order to understand the phenomena which have marked the evolution of the Universe, we need to observe the sky in the whole range of electromagnetic waves, hence from space. It also nicely illustrates the evolution of our Universe as well as the main contribution of space techniques to the discovery and characterisation of the main features of cosmological significance such as the cosmic microwave background, the recently discovered cosmic infrared and optical backgrounds emitted mainly by warm dust and normal stars respectively, and the cosmic X-ray background discovered in 1962 and in the interpretation of which Lo Woltjer made such important contributions.

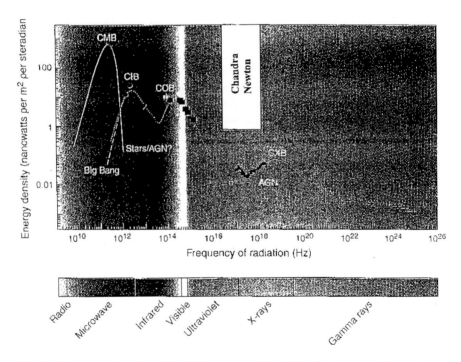

Fig. 1. The energy density of the Universe over the whole electromagnetic spectrum (Hasinger 2000)

Let me discuss now four of the most essential contributions of space astronomy.

3 High Energy Astronomy

Table 1 lists the most important space observatories of the – not yet finished – space era, with the exception of the high-energy telescopes or experiments which are themselves detailed in Table 2. The direct comparison between the two tables strikingly illustrates the enormous efforts devoted to high-energy astronomy, a consequence of its unique potential of discoveries and of its enormous diagnostic power. It also illustrates the very fast progress of technologies witnessed in this domain, in particular reflecting optics and telescopes, detection techniques and multidimensional arrays of detectors.

We should remember the first observation of Giacconi et al. (1962) of an X-ray emission observed outside the Earth's atmosphere (Fig. 2). When we compare this first result, a "première" at that time, with the best images now obtained with the Chandra Telescope, we can measure the progress which has been accomplished over the past 40 years. However, astronomy is not only based on images. The power of X-ray spectroscopy has been amply demonstrated in the course of this meeting, in particular in revealing the presence of black holes through their very characteristic line profiles, showing the intense gravitational red shift induced by these very massive objects (Cagnoni et al. 1998).

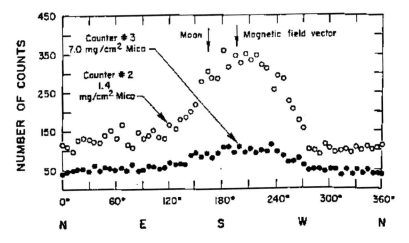

Fig. 2. The historical observation by Giacconi et al. of an X-ray emission from the direction of the galactic centre

The availability of excellent quality spectra, obtained by ESA's XMM-Newton in particular (Jansen 1999), adds an extremely powerful diagnostic capability to our means of investigation. This is strikingly illustrated in Fig. 3, which shows the spectrum of HR 1099. We can assess the quality of the data when we realise that the noise level is smaller than the thickness of the line.

Table 1. Main (non solar) astronomical satellites launched or in preparation, excluding X-ray satellites

	RADIO
1997	VSOP (J)

	INFRARED
1983	IRAS (US-NL-UK)
1989	COBE (US)
1995	ISO (ESA + J)

	VISIBLE
1989	HIPPARCOS (ESA)
1990	HUBBLE (US + ESA)

	ULTRA-VIOLET
1968	OAO-2 (US)
1972	TD-1 (ESRO)
1973	ANS (NL + US)
1973	OAO-3 (US)
1978	IUE (US + UK + ESA)
1992	EUVE (US)
1999	FUSE (US + EUR)

	GAMMA
1972	SAS-2 (US)
1975	COS-B (ESA)
1990	SIGMA (F + RUS)
1991	CGRO (US)
2001	INTEGRAL (ESA + RUS)

Table 2. Main X-ray astronomy satellites and experiments launched or in preparation

Mission Name	Instrumentation	Year	Imaging					
			Range (keV) E-Low	E-High	Resolution (")	Eff.Area (cm^2)	Energy Res. E/delta-E	FOV (')
Vela-5B		1969						
Uhuru	PC	1970						
OSO-7	PC	1971						
Copernicus	PC, M	1972						
ANS	PC, X	1974						
UK-5	PC	1974						
COS-B	PC	1975						
SAS-3	PC	1975						
OSO-8	PC, X	1975						
HEAO-1	PC	1977						
Hakucho	PC	1979						
HEAO-2	PC+IT, MCP, TG, X, SSS	1978	0.25	3.00	2.00	20.00	0.90	25.00
UK-6	PC, M	1979						
Tenma	GSPC	1981						
Astron		1983						
EXOSAT	IT+PC/MCP, TG, PC, GSPC	1983	0.05	2.00	18.00	10.00	0.90	120.00
Ginga	PC	1987						
Gamma-1	PC							
Granat	PC	1989						
ROSAT	IT+ PC/MCP	1990	0.10	2.50	20.00	240.00	5.00	120.00
ASCA	IT+GSPC/SSS/CCD	1993	0.40	12.00	30.00	105.00	40.00	22.00
RXTE	PC	1995						
BeppoSax	IT+GSPC	1996	1.30	10.00	75.00	150.00	16.00	56.00
Abrixas	IT+CCD	1999						
Chandra	IT/TG+CCD/MCP	1999	0.10	10.00	0.50	600.00	40.00	16.00
XMM-Newton	IT/RGS+CCD,OM	1999	0.10	15.00	4.00	2600.00	40.00	30.00
Astro-E	IT+CCD/Cal	2000						
Spectrum-X	IT+GSPC/SSS/CCD/X/PC							
Constellation-X	IT+Cal	2008	0.10	10.00	10.00	15000.00	1000.00	2.50
XEUS-1	IT+TED, APD	2013	0.10	10.00	2.00	40000.00	1000.00	5.00
XEUS-2	IT+TED, APD	2013	0.10	10.00	2.00	200000.00	1000.00	5.00

PC	Proportional Counter
M	Non-imaging telescope
X	Bragg Crystal
IT	Imagine Telescope
MCP	Multi-Channel plate
TG	Transmission grating
SSS	Solid state spectrometer
GSPC	Gas-scintillation PC
RG	Reflection Grating
CCD	Charge Coupled Device
Cal	Calirometer/Bolometer
TED	Transition Edge Detector
APD	Active Pixel Device

Until recently, such quality spectra were available only for the Sun. The richness of the details seen in Fig. 3 opens up new possibilities of measuring temperatures and densities in the atmospheres of a very large number of stars and nebulae.

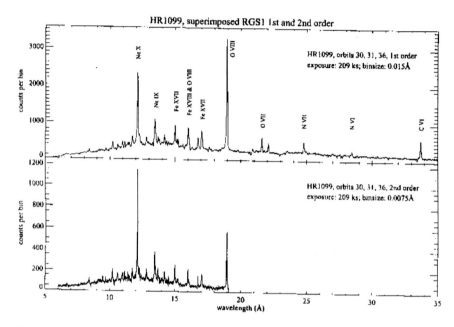

Fig. 3. Spectrum of HR 1099 obtained by the Reflection Grating Spectrometer on board ESA's XMM-Newton satellite

4 The Possibility of Achieving the Highest Possible Angular Resolution

The competition between ground-based and space-based telescopes should not be denied. Neither should it be overlooked. Even though a slight advantage for space seems to emerge, with of course the noticeable role played by the Hubble Space Telescope, the coming of age of ESO's VLT, whose existence we owe to Lo's vision, demonstrates that the difference between space and ground astronomy for the visible is not very large. The costs of images of comparable qualities would certainly favour ground-based techniques, illustrating once more the principle that we should not do in space what can be done as well and much cheaper from the ground. However, the days when space observatories will remain definitively unbeaten are not far off. We will come back to this issue at the end of the paper. This is obviously the case in the two extreme parts of the spectrum.

In the X-rays, Chandra with its one arc-sec resolution, although far away from the performance of the VLT and HST, provides the best images to date in this domain. However, the possibility of building X-ray interferometers with resolutions of a thousandth of an arc-sec does not seem to be too remote a dream. Concepts of such instruments are now under study in the USA and in Europe.

Gamma-ray astronomy is only possible from space but there, high angular resolution is far remote. The best we can hope for at this moment is to reach the range of a few arc-minutes. Spectacular progress has been witnessed in the recent past, in particular with the EGRET instrument on NASA's Compton Gamma Ray Observatory (CGRO) which has just discovered 170 new and unknown sources at 100 MeV, most likely in the Milky Way (Gehrels et al. 2000).

More progress will come in the future with ESA's Integral mission (in cooperation with Russia) to be launched in 2002, and NASA's Gamma-ray Large Area Space Telescope (GLAST) to be launched in 2005. Coded masks and bi-dimensional detectors now offer real imaging capabilities. However, much more progress is necessary to bring gamma-ray astronomy to the level of X-ray and *a fortiori* visible astronomy.

In the radio range, of course, space is the ultimate place to conduct VLBI observations: on the ground, the dimensions of the Earth itself impose limitations on the baselines of the interferometers. The Japanese have clearly shown the way with their VSOP mission and it is quite likely that in the course of the next century space VLBI will develop and open up astronomical imagery with resolutions in the micro arc second range.

5 Infrared Astronomy

Table 1 strikingly shows the infantile status of infrared astronomy with only three missions launched as of now: IRAS (USA-NL-UK), COBE (USA) and ISO (ESA-ISAS). This certainly does not pay tribute to the richness of this spectral domain and only reflects the real technical difficulties encountered in particular in the area of cryogenics and of detectors.

While the near infrared up to a few μm is still within reach of HST, the medium and certainly the far infrared require cooled optics and detectors. By the very nature of their longer wavelengths, infrared photons are less sharply imaged than those in the visible and without interferometry, observations will perform less well than in the visible. So far, the best observations in the far infrared have been provided by ESA's Infrared Space Observatory, ISO, which was operated as a real observatory after its launch in November 1995 for 28 months, nearly a year more than its nominal lifetime. Lo Woltjer chaired the ISO Observing Time Allocation Committee.

ISO has demonstrated unequivocally the diagnostic power of the infrared in identifying the main components of the Cold Universe and in particular the composition of interstellar molecular clouds and of dust which, contrary to the visible, does not block the infrared, thereby revealing new objects in the central parts of galaxies or in the regions where stars are formed. The universal presence of water, in the form of either vapour or ice, is one of the most spectacular outcomes of ISO. Its rich harvest has paved the way to the science exploitation of NASA's long awaited SIRTF mission.

The infrared is well adapted to the development of interferometers in space. Such instruments will probably be the major source of discovery of terrestrial planets and this for two reasons. First, the contrast between the planet and its mother star is higher in the visible than in the infrared, and second the spectral signature of the ingredients of life (H_2O), or of the presence of life (O_2, O_3) lie in the mid-infrared (Fig. 4). Given these prospects, both ESA and NASA are keen to study different concepts of infrared interferometers: IRSI-DARWIN (ESA) and the Terrestrial Planet Finder (NASA). However, the ambitions and the dimensions of these instruments clearly demand that these projects be conducted through international cooperation, and both agencies are already joining forces in the studies of their respective concepts.

6 In Situ Exploration of the Universe

The present state of space techniques does not allow in situ exploration beyond the Solar System. Nevertheless, the numerous interplanetary probes and orbiters launched by the Americans, the Soviets, the Europeans and the Japanese to study the Sun and its planets have drastically improved with time and have changed our perception of the Solar System and how it was formed. The power of high-resolution imagery was demonstrated by the Soviet and American missions with a climax reached with the most recent pictures of Mars.

Several tens of new moons around Jupiter, Saturn and the other external planets have been discovered. Volcanism is no longer a terrestrial phenomenon only, but has also been found to exist on Mars, Venus and Io. Meteoritic bombardment has been very intense throughout the Solar System and has led to some spectacular scars on Earth and on all solid bodies including asteroids.

The presence of water has been confirmed by ISO in the atmosphere of Jupiter, Saturn and Titan on the surface of which ESA's Huygens probe will land in 2004, and the Soviet Vega and the ESA Giotto missions to Halley's Comet in 1986 measured vast quantities of water in the nucleus and basically confirmed the "dirty snow ball" model (Mendis 1987).

In the period 1993–1997, Lo Woltjer was chairing the SSAC, a high level Committee which advises the ESA Director General on space science matters. This four-year period is unusually long since the term of the SSAC chairman

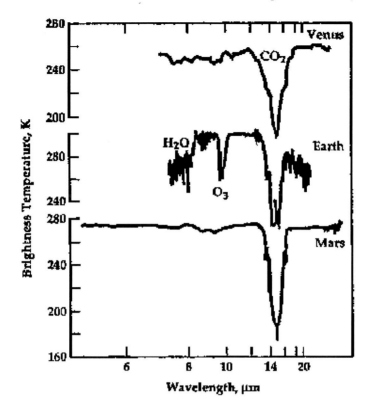

Fig. 4. Comparison of the mid-infrared spectra of Venus, the Earth and Mars, evidencing the dominance of water and ozone as indicators of the presence of life on a planet (source: Jet Propulsion Laboratory)

is for two years with the possibility of a one-year extension! However, Lo's excellent chairmanship fully justified that exception.

In 1994, I requested the SSAC to prepare a 10-year plan of space science missions following the launch of FIRST, foreseen in 2007, which would signal the accomplishment of ESA's Horizon 2000 long-term plan. Several experts joined the SSAC and formed the Survey Committee, which met in Rome in October 1994 to formulate what would become the Horizon-2000 Plus Programme.

If the race to the Moon in the early years of space research was a political challenge, the race to Mars is in addition a scientific as well as a public information challenge. The appeal of the red planet was clearly present in Lo's mind at the time of the Rome meeting, and he insisted that a Mars mission should be a priority for ESA and even a cornerstone of Horizon 2000 Plus.

The Survey Committee, for opportunistic reasons, did not follow that recommendation and selected rather a cornerstone mission to Mercury. However, when reviewing the implementation of Horizon 2000 Plus at the January 1997 SSAC meeting, Lo requested that a small task force be set up to investigate whether a low cost mission to Mars would be possible as a new element in the programme and a way of implementing the priority for Mars. Such a possibility would in addition offer the planetary community an opportunity to rapidly replace the science which had just been lost in the failure of the Russian Mars 96 mission. The task force came to a positive conclusion and the SSAC fully endorsed the concept of the Mars Express mission. Final approval by ESA's SPC was obtained in 1998 and Mars Express will be launched in 2003 at a cost of 166 million Euros, making it the cheapest Mars mission in the world.

That decision did not please everybody: the astronomers had to accept a delay of the Planck Surveyor mission, which was unavoidable anyway because of the difficulties foreseen in the development of its very demanding payload. Remarkable however, was the fact that Lo himself, an astronomer, strongly supported it!

Space techniques have also allowed us to study the Sun in detail, in particular thanks to the ESA-NASA SOHO mission and the techniques of helioseismology. Such studies can certainly be – and have been – done from the ground but space-borne instruments provide the unique asset of absolute continuity in the observations and furthermore are free from any perturbations created by the Earth's atmosphere, thereby reducing the noise by an order of magnitude (Gabriel et al. 1998). Measurements from space also offer a unique advantage for the measurement of the Solar Constant and of its variations with time (Fig. 5).

These two last examples allow me to raise again the basic question: is it necessary to conduct visible astronomy from space?

7 Is Space Astronomy in the Visible Worthwhile?

The answer is yes ultimately!

The Next Generation Space Telescope, dear to the heart of Lo, will provide unparalleled spatial resolution in particular in the near infrared, and space interferometry as mentioned earlier will provide baselines beyond the limits which can be reached from the surface of the Earth.

Without space techniques, milli arc-sec astrometry would have remained a nearly impossible challenge for the astronomers. Hipparcos has created a genuine revolution, not only in astrometry but also in astrophysics, with accuracies of one to two milli arc-secs, and the unique advantage of making global measurements with the same instrument over the whole celestial sphere. A revolution indeed not only because of this quantum jump in resolution, but also because of the fact that a new family of astrometry missions has

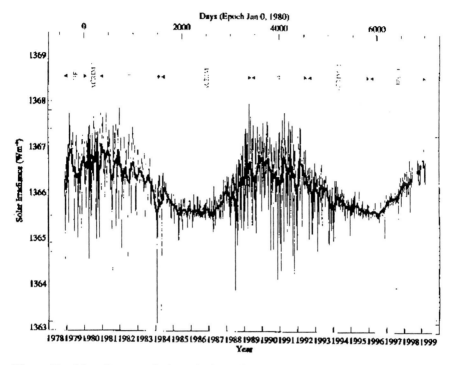

Fig. 5. Total irradiance variations during solar cycles 21–23 as recorded by several satellites since 1978 (kindly provided by C. Fröhlich)

now been born, both in the USA as well as in Europe, with in particular the GAIA mission proposed as a cornerstone of Horizon 2000 Plus (Gilmore et al. 2000). GAIA aims at a resolution of a few micro arc-seconds, two to three orders of magnitude better than Hipparcos, and for a billion stars, allowing in particular to observe several tens of thousands of new planets.

The unique advantage of the global nature of Hipparcos will certainly be fully exploited by ESA's Planck Surveyor mission in its observations of the temperature fluctuations of the cosmic background where high angular resolution is required to determine the key cosmological parameters of the Universe. Here satellites offer a real advantage over ground-based and balloon borne instruments.

8 Conclusion

To explore the whole information content of the electromagnetic spectrum, space astronomy is absolutely essential. However, the costs of space missions are very much higher than those of most of even the larger ground-based observatories and the principle that "what can be done from the ground should not be done in space" should be implemented vigorously.

Nevertheless, how can we justify these expenditures? For the man in the street, space observation and its unique potential of discoveries still exerts a fascination especially among young people and also among ordinary people. This is unique and we should not deprive the taxpayers of this source of dreams.

But there is something more that space astronomy offers to society. This is the capability of putting to work the most talented scientists, engineers and technicians. More than 30 000 people are involved in Europe in the definition, development and exploitation of space missions. The rigour and the high level requirements of space projects where failure is not an option, force them to work to the best of their capabilities and of their talents.

This symposium has been a unique opportunity to celebrate one of the strongest personalities of the astronomy community, Lo Woltjer and his vision which extends far beyond the average horizon. Let us hope that he will advise and lead us for a long time in the future.

Lo, let me thank you for everything you have done for space science and again let me wish you a very happy birthday!

References

1. Cagnoni, I., Della Ceca, R. and Maccacaro T., 1998, Astrophys. J. 493, 54–61
2. Fröhlich, C., 1988, ESA SP-418, 7–10
3. Gabriel, A.H., Turck-Chieze, S., Garcia, R.A., Pallé, P.L., Boumier, P., Thiéry, S., Grec, G., Ulrich, R.K., Bertello, L., Roca-Cortès, T. and Robillot, J.M., 1998, ESA SP-418, 61–66
4. Gehrels, N., Macomb, D.J., Bertsch, D.L., Thomson, D.J. and Hartman, R.C., 2000, Nature 404, 363–365
5. Giacconi, R., Gursky, H., Paolini, F. and Rossi, B., 1962, Phys. Rev. Letters 9, 439–443
6. Gilmore, G., de Boer, K., Favata, F., Hoeg, E., Lattanzi, M., Lindegren, L., Luré, X., Mignard, F., Perryman, M. and de Zeeuw, P.T., 2000, Proceedings of SPIE 4013, in press, March 2000, Munich, J.B. Brekinridge and P. Jakobsen editors
7. Hasinger, G., 2000, Nature 404, 443–446
8. Jansen, F.A., 1999, ESA Bulletin 100, 9–12
9. Mendis, D.A., 1987, Astron. and Astrophys. 187, no. 2, 939–948

After-Dinner Speech in Honour of Lodewijk Woltjer

Reimar Lüst

Max-Planck-Institut für Meteorologie, Bundesstrasse 55, D-20146 Hamburg, Germany

I accepted the invitation to speak on this occasion with great pleasure, since it gives me a chance to express my respect, admiration and also my gratitude for our longstanding, truthful connection for Lo Woltjer in front of a large and very competent audience. But I must confess that I am also somewhat hesitant, since Lo Woltjer is certainly not a very simple person. There are many facets of him which should be brought up in such a talk.

But how could I cope with this task in an after-dinner speech which must fulfil the two conditions of a dress of a young lady, namely "as short as possible but covering all the essentials".

In order to achieve this I shall concentrate on three characterizations of Lo Woltjer.

1. He is an outstanding scientist
2. He is an excellent manager
3. He is a perfect gentleman

Let me try to cover these three essentials knowing that not everything is covered, but this sometimes also applies to the dress of a girl which makes it even more attractive.

When I looked into his curriculum vitae I found that our two lives developed somewhat in the same pattern and our ways crossed each other from time to time.

We both started very actively in science, attacking new problems in particular in Astrophysics and Plasmaphysics. But as time passed, he and I had to accept certain responsibilities. And finally we had to reduce our scientific work and accepted to be managers for European organisations.

The first important responsibility of Lo Woltjer was the chairmanship of the Astronomy Department at Columbia University in New York. For eight years, from 1964 to 1974, he held this position. In addition, Lo Woltjer had the difficult task of Editor of the Astronomical Journal.

In 1975 he was ready to start as a real manager, as the Director General of the European Southern Observatory (ESO), while I had made this change already in 1972 as President of the Max Planck Society. He served for 12 years at ESO and I for 12 years at the MPI Society.

Our paths crossed for the first time in Princeton in 1959. The two houses of the Institute of Advanced Studies in which we lived were adjacent in Einstein Road 46 and 48. I will come back to this period.

The second crossing point was Garching, when Lo Woltjer moved into the new headquarters of ESO. But before that he was already a member of the Visiting Committee for the Max Planck Institute of Astrophysics in Munich, later for the MPI of Radioastronomy in Bonn, and finally for the MPI of Astronomy in Heidelberg where he is presently still the chairman. And finally he conducted a review of the Space Science Department, when I was in Paris responsible for ESA.

1 The Outstanding Scientist

His scientific achievements are impressive, as well as the number of his publications. Altogether there are 155 printed publications which cover a wide field in Astrophysics.

I hope I do not bother our non-scientist companions too much when I just read the titles of the first publications. He started as a real astronomer: The first publication was an investigation of the spicules and the structure of the chromosphere. The following publications dealt with radial velocities of stars. I read only the first one when I was a Fulbright Fellow at the Princeton Observatory working with Martin Schwarzschild.

But his paper on the Crab nebula made the name of Lo Woltjer known in the astrophysical community. At that time I was in a group of Ludwig Biermann in Göttingen where we were interested in the new fields of plasmaphysics and magneto-hydrodynamics. I remember how excited we were when Oort and Walraven published their results on the polarisation of the light from the Crab nebula, based on photographs taken by Baade in 1942. This was the first real demonstration of magnetic fields outside the Solar System.

Lo Woltjer in his Ph.D. thesis analyzed in detail the structure of the magnetic field in the Crab nebula. (I do not know how I could explain the importance of force-free fields to the non-astronomers present here.) We were both working on this problem at that time, but we did not know each other.

I could continue to mention many subjects in Astrophysics which Lo Woltjer had dealt with in great competence. When he became older he was very often asked to have the last word at a meeting. Therefore his list of publications includes a number of papers with the simple title: "Concluding Remarks".

2 The Excellent Manager

This was also his strength when he acted as manager. After difficult discussions he could bring the matter to the point by making clear that this was the last word. Of course I had never directly experienced how he handled and managed ESO as a Director General.

But I got reports on how he acted as a chairman of the Space Science Advisory Committee of ESA. This is a very influential committee in ESA, and Lo Woltjer served as a chairman for an unusually long period of four years and not only for two years.

A very competent observer described the procedure at such meetings as follows: "He would let people talk. Not for long, however. Then he would raise his right hand to shoulder height, extended, parallel to his side. Then he would flex the wrist and bring the hand down slowly like a guillotine. When the hand was at ten centimeters from the table, the speaker had better be finished with whatever he had to say. Nothing was ever said about this procedure, but people learned it fast. Some SSAC members probably still have nightmares about it. The discussion could be long and complex. At the end one had to put a recommendation into words and at this point Lo Woltjer would say: I have tried to put into words the recommendation, and I have modified it according to your wishes during the discussion. Or something like that. Then he would produce a one-page sheet handwritten in green ink, which he had obviously prepared some days before. Everybody was happy and congratulating himself on a good job done, but normally there were very few corrections (or none) to the original text."

A visitor described how he observed Lo Woltjer acting as DG of ESO. "I will never forget my first visit to him at ESO Headquarters in Garching. His secretary, Ulla (now his wife) opened the door of his office. I discovered it: empty with a bunch of yellow flowers: the same yellow as his tie (he must have only one). The desk was as clear as the desert. Behind the desk was the frightening Woltjer reading the Herald Tribune! There was the DG of ESO."

His achievements in ESO as DG are certainly outstanding. Let me indicate only three.

1. In 1982 he succeeded to convince two countries to become members of ESO: Switzerland and Italy.

2. He made sure that the ESO Council did not use the entrance fee of Italy and Switzerland to reduce the contributions of the older members which would have been the normal reaction of the governments, but that they used the entrance fee to build the New Technology Telescope (NTT).

3. Under his leadership the ESO Council decided to build the Very Large Telescope (VLT) and he was responsible that a new location in Chile in addition to La Silla was found and bought.

But he acted always in a very quiet way and was never hectic.

3 The Perfect Gentleman

This brings me up to the third point since a perfect gentleman is never hectic. I am sure that all the ladies who know him will agree that he is the best dressed Astronomer.

When we were together in Princeton his wife told me: You know Lo is a snob. I could not agree with this statement since I liked already his lifestyle.

Certainly he loves good living. This can be demonstrated also by the places where he chose to live.

From Leiden he went to the United States. There he started at Yerkes Observatory near Chicago. From there he moved to Princeton which is not too far from New York, but finally he moved to New York where he lived for 10 years.

From there he moved to Geneva. Lo, I want to apologize that I was certainly responsible that you could not stay forever in Geneva as you would have preferred but that you had to move to this very odd place Garching near Munich.

In 1973 I was asked by Bengt Strömgren, the chairman of the ESO Council at that time, if I could discuss together with him and Blaauw, at that time the DG of ESO, a problem created by the German delegation. In Copenhagen we had an intensive discussion. I was not so surprised about the difficulties with the German delegation since normally either the German or the French delegation are difficult. At that time the German government was very upset since ESO headquarters had moved practically to Geneva, while according to the convention Hamburg was the official seat.

So finally we came to the conclusion that it would not be wise to change a convention but to find another place in Germany since Hamburg-Bergedorf was no longer considered suitable. I mentioned Garching in a very selfish way, since I wanted to move the Max-Planck-Institute for Astrophysics from Munich to Garching. The members of this Institute did not like this. But if ESO would move to Garching I had a very good argument.

So, Lo, I apologize that in 1980 you had finally to move to Garching, where, in 1981, the new buildings were inaugurated. But I know your preference for Switzerland and Italy. Certainly this was the deeper reason you managed that Italy and Switzerland became members of ESO. Since then you always had a good excuse to continue your lifestyle in Switzerland and Italy. The Symposium is a demonstration of this. The Symposium is held here in Italy while your permanent home is in Switzerland.

In concluding my speech let me thank you for what you have done in Astronomy and for the Astronomers and also for your friendship. We all hope you will give advice to the astronomical community for many more years and we wish you a long continuation of your lifestyle as a perfect gentleman.